Social Class, Social Action, and Education

Social Class, Social Action, and Education

The Failure of Progressive Democracy

Aaron Schutz

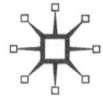

SOCIAL CLASS, SOCIAL ACTION, AND EDUCATION
Copyright © Aaron Schutz, 2010.
Softcover reprint of the hardcover 1st edition 2010 978-0-230-10591-1

All rights reserved.

First published in 2010 by PALGRAVE MACMILLAN® in the United States—a division of St. Martin's Press LLC, 175 Fifth Avenue, New York, NY 10010.

Where this book is distributed in the UK, Europe and the rest of the world, this is by Palgrave Macmillan, a division of Macmillan Publishers Limited, registered in England, company number 785998, of Houndmills, Basingstoke, Hampshire RG21 6XS.

Palgrave Macmillan is the global academic imprint of the above companies and has companies and representatives throughout the world.

Palgrave® and Macmillan® are registered trademarks in the United States, the United Kingdom, Europe and other countries.

ISBN 978-1-349-29020-8   ISBN 978-0-230-11357-2 (eBook)
DOI 10.1057/9780230113572

Schutz, Aaron.
    Social class, social action, and education: The failure of progressive democracy / Aaron Schutz.
        p. cm.
    1. Education—Social aspects—United States. 2. Democracy and education—United States. I. Title.

LC191.4.S38 2010
370.11'509730904—dc22   2010027992

A catalogue record of the book is available from the British Library.

Design by Scribe Inc.

First edition: September 2010

10 9 8 7 6 5 4 3 2 1

Transferred to Digital Printing in 2011

To my mother, who was, in her own idiosyncratic way, a true scholar.

She always got mad when she read my papers for school.

"You just banged this out, and you're going to get an A," she would complain. "You could do so much better if you took the time to actually think about what you are saying."

I didn't bang this one out, Mom.

Listen to me, college boy, you can
keep your museums and poetry and string quartets
'cause there's nothing more beautiful than
[power] line work.

—Todd Jailer, "Bill Hastings"

Contents

Acknowledgments — ix
Introduction — 1

Part I   Overview
1 Social Class and Social Action — 9

Part II   Collaborative Progressivism
2 John Dewey's Conundrum: Can Democratic Schools Empower? — 51
3 John Dewey and a "Paradox of Size":
   Faith at the Limits of Experience — 71

Part III   Personalist Progressivism
4 The Lost Vision of 1920s Personalists — 91
5 The Free Schools Movement — 127

Part IV   Democratic Solidarity
6 Community Organizing: A Working-Class
   Approach to Democratic Empowerment — 161

Part V   Case Study
7 Social Class and Social Action in the Civil Rights Movement — 189

Part VI   Conclusion
8 Building Bridges? — 221

Notes — 229
Index — 263

Acknowledgments

This book represents more than a decade of work, of following my nose through unexpected turns and twists. Over this time, I collected debts to those who helped me find my way. At the University of Michigan, my most important mentor was Pamela Moss, who taught me what it means to be a professional in the scholarly arena. I would not have accomplished near as much without her support and friendship. Jay Robinson took a chance on a candidate with an odd background for their PhD program and trusted that I would find my way, somehow. And Anne Gere gave me the space I needed to find that way. Larry Berlin gave encouragement and the model of his own strong commitment to old-fashioned scholarship. Others who provided insight or read portions of this work and gave helpful comments include Todd DeStigter, David Granger, Craig Cunningham, Tobin Siebers, Michael Fielding, Kathleen Knight Abowitz, Larry Marx, David Liners, and Jim Garrison. I also want to thank my coparticipants in Milwaukee Innercity Congregations Allied for Hope (MICAH) and the students in my Introduction to Community Organizing. Dee Russell probably forgets that he read an early version of Chapter 2 from a distance and asked whether I really thought Dewey was as stupid as I had made him out to be (framed much more kindly than that, of course). I have kept this question in mind ever since. My brother, Austin, often challenges me with his idealism, providing a useful counter to my ingrained cynicism.

I am sure I have forgotten others.

These critics and supporters are, of course, fully responsible for any mistakes that remain in the current manuscript.

The work of a number of other scholars, most of whom I have not met, was extremely influential on the development of this book. Michael McGerr and Sheldon Stromquist's recent books on turn-of-the century progressivism and its middle-class character were invaluable. Annette Lareau's deeply insightful study of class-based parenting practices in the United States was a crucial foundation for my larger cultural argument. Paul Lichterman's and Fred Rose's books on the relationships between social class and social action practices also provided critical empirical grounding for the overall argument of this book. Cornel West's brief acknowledgment that Dewey's philosophy was middle class in his 1989 *American Evasion of Philosophy*, which I read as a doctoral student, helped assure me that I was on the right path at an early point in my career. It turns out that Walter Feinberg made some similar arguments about the class basis of Dewey's perspective (going further than I do in some ways) more than thirty years ago in his *Reason and Rhetoric*, although I did not realize this until he pointed it out to me. It goes without saying that Robert Westbrook's pathbreaking biography of John Dewey was a critical support. The chapter on the "paradox of size" is deeply indebted to the work of Jane Mansbridge, who has done some of the most insightful writing about democracy over the last few decades. If Casey Blake had not done such careful scholarship in his book *Beloved Community* on the difficult-to-capture vision of the Young Americans, the chapter on the 1920s would likely not have been written. Similarly, without Ron Miller's insightful *Free Schools, Free People*, I likely never would have made the connection between progressivism and what I call the "personalists" of the 1960s and 1970s. Sanford Horwitt's biography was a vital window into the complex life of Saul Alinsky. Barbara Ransby's biography of Ella Baker and Lance Hill's book on Deacons for Defense, were key pillars on which my case study rests, along with more familiar classics on the civil rights movement by Clayborne Carson and others.

I would also like to acknowledge the work of Michael Apple and Richard Brosio who, in their different ways, tried to keep the challenges of social class on the radar of educational scholars during the many years when class was unfashionable.

Reaching further back in my career, Maxine Greene's *Dialectic of Freedom* first brought me to the question of public action. In fact, to some extent, my dissertation was an exercise in following back her reference list. And the spirit of Hannah Arendt, discovered through Greene, haunts these pages in ways I cannot ever know, although I have mostly left Arendt behind.[1]

This volume could not have been written without the support and love of my family: my wife, Jessica, and my daughters, Hiwot and Sheta, who, when asked what daddy loves best, answer "being on the computer."

Introduction

The term "progressive" returned with a vengeance during the first decade of the twenty-first century.* With "liberal" under attack, the left turned back to a name that had rallied champions of social transformation throughout the first half of the prior century. Of course, most of those who call themselves "progressives" today are not referring to anything particularly specific—it has largely become a vague collective reference for a wide range of left-leaning groups. But the increasing use of the term has increased interest in progressivism as a more substantive concept and social vision.

This volume focuses on a fairly narrow aspect of progressivism: its conceptions of democracy. I trace how two understandings of progressive democratic practice emerged in the early decades of the twentieth century that I call "collaborative" and "personalist." And I show how these visions of "authentic" democracy still deeply influence our ideas about social justice and education in America.

"Collaborative" progressivism developed as a coherent perspective at the end of the nineteenth century among a loosely connected group of middle-class progressives—religious leaders, scholars, and activists. Together, this group imagined a world in which bureaucracy and elite control would slowly dissolve into a flat, truly collaborative, and egalitarian society. If people would only work together, they believed, they could solve the growing problems of poverty and inequality in an increasingly industrial society. The collaborative progressives understood that America was far from their ideal, and most were realistic enough to understand that their full utopian vision was probably unachievable. Nonetheless, they threw themselves into a wide range of efforts to bring about the conditions necessary to achieve as much as they could. The most sophisticated theorist of this democratic ideal, as I discuss in Chapters 2 and 3, was John Dewey. In its general outlines, however, the collaborative vision differed little across the broad range of progressive intellectuals.

---

*Except where they add something to the arguments made later in this volume, I leave citations to the more substantive chapters that follow.

Some decades after the emergence of collaborative progressivism, during the "gay" twenties and later in the 1960s, as I describe in Chapters 4 and 5, another vision of holistic democracy coalesced among a different group of progressives that I call the personalists. This group has largely been forgotten in the academic literature, especially in education. Unlike the collaborative progressives, who sought pragmatic strategies for fixing a society rife with inequality and social conflict, the personalists came of age during times of relative prosperity, when it seemed likely (to them) that poverty and discrimination could simply disappear. At these moments it seemed reasonable, for the relatively privileged in society at least, to leave many of collaborative progressives' social concerns behind. Instead of developing practices for communal problem solving, they envisioned egalitarian communities where authentic relationships would nuture each member's distinctive personality. The personalists sought to release the capacities of unique individuals, looking to romantic ideals of creative, fully embodied, and emotionally free people. As I explain in more detail later, the term "personalist" seems to fit this group best because of their combination of communal *and* individual aims.

On first glance, the personalist ideal of democracy can seem quite different from the apparently more sober vision of the collaborative democrats. In fact, however, the overall social aims of both groups were quite similar. As I show in Chapter 5, the core assumptions about human nature that informed both were much the same. The collaborative progressives focused on the challenges of effective cooperation. The personalist progressives focused on the release of the capacities of unique individuals. And each side criticized the other for its excesses—the collaboratives attacking personalists for their lack of a concrete vision of joint action and the personalists attacking the collaboratives for their failure to fully appreciate the importance of individual freedom and authentic human relationships. But both nonetheless acknowledged and emphasized the importance of both aims. More generally, both camps sought to foster a new, more freely dialogic, and less hierarchical society. The collaboratives and the personalists, therefore, lie on a common continuum of "democratic" progressive thought.

These democratic ideals have remained compelling for a broad range of progressive intellectuals into the twenty-first century, even though they have proved extremely difficult to enact in actual practice. Why? The answer, I argue, lies largely among progressives themselves, among whom I count myself as a member (albeit a critical one). Scholars, especially in education, find collaborative and personalist visions of democracy compelling because they reflect advanced versions of the cultural practices most familiar to the vast majority of us in our families, schools, business dealings, and associations. The dreams of progressive democracy are literally embodied within the selves and social institutions of intellectuals in America. In other words, we like Dewey at least in part because Dewey was like us.

The central influence on our long romance with progressivism, I argue, has been middle-class culture. The book begins in Chapter 1, therefore, with an analysis of the emergence of the middle and working classes in the United States. I show how the middle class slowly split as a group from the working class over the last half of the nineteenth century and how progressivism emerged in parallel with an increasingly distinct middle-class professional culture. Chapter 1 lays out key characteristics of each class's cultural lifeways, drawing together research describing relationships between class cultures and social action practices in America.

Progressives of all stripes have always shied away from models of democracy drawn from the experiences of other classes. This has been especially true of models emerging out of the working class, which, from a progressive perspective, have often seemed brutish and primitive. Progressives rejected working-class tendencies to emphasize the inevitability of aggressive social conflict. And progressives were uninterested in the practical demands of mass solidarity reflected in the strategies of labor unions and, more recently, community organizing groups.

In fact, the "backwardness" of working-class culture was perceived from the beginning by progressives as a core social barrier to the achievement of authentic democracy. Many progressive intellectuals struggled in their writings with how to "uplift" the working class. They sought to develop pedagogies, for example, that might initiate these "others" into adequate capacities for democratic citizenship. Even the personalists—who often looked to more "primitive" cultures for alternatives to the banality of modern middle-class life—were repelled by the lack of focus on individual actualization and aesthetic expression among the lower classes. In fact, a third major group of progressives, "administrative" progressives, argued that broad-based democracy was an impossibility in the modern world in no small part because of the seemingly unredeemable ignorance of the working classes.

Of course, social class was not the only source of progressive discrimination. Racism was an ongoing factor as well. In this volume, however, I limit my focus to the ways that progressive racism emerged out of concerns about social class.[1] The racism of many early progressives emerged in large part out of their broader arguments about the backwardness of less "advanced" cultures, leading to judgments, for example, about what they saw as the especially deficient nature of African American culture.[2]

The collaborative progressives of the first half of the twentieth century were interested in more than democracy. They also sought to combat corruption and address key social inequalities. They attempted to "rationalize" a chaotic society, looking to science as a savior. And, with the administrative progressives, they accomplished many important social goals, including the creation of unemployment insurance, child labor laws, new voting regulations, the Food and Drug Administration, and social security, among many others.

The larger hopes of collaborative and personalist progressives for a more democratic society, however, met almost complete failure.[3] It was instead the antidemocratic vision of administrative progressives that ultimately had the most impact on the social structure of American society, creating the public and private bureaucracies that still manage much of our lives today. The efforts of the collaboratives and personalists to foster their vision of democracy remained mostly limited to voluminous writings, experiments in a few schools and other contexts, and largely ineffective political interventions. In contrast, while they may not have achieved the kind of benevolent society they desired, the administrative progressives, were nonetheless extremely successful in intensifying the centralization of many government and other institutions' functions under the control of a professionally guided bureaucracy.

This book focuses on the educational component of progressivism, in part because collaborative and personalist conceptions of democracy have remained more influential in education than elsewhere. With respect to collaborative democracy, this is largely the result of the continuing dominance in the field of John Dewey's extensive writings on pedagogy and learning. In the academic literature in education it is nearly impossible to find writings on democratic education that do not embody key aspects of his vision, even when Dewey himself is not explicitly mentioned. The personalist ideal is, if anything, even more influential, albeit in more diffuse ways, among educators and educational scholars, even though the key writers and pedagogues that best formulated this vision—Margaret Naumburg, Caroline Pratt, Paul Goodman, and others—are largely forgotten. Core aspects of the personalist vision live on, for example, in the popularity of Nel Noddings's formulation of "caring" schools.[4]

This book is not only written for educational scholars, however. As a case study, the arena of education provides a useful example of patterns visible in discussions about democracy across the social sciences and humanities. In these other fields, as well, one will find among those who cherish democracy a deep preference for aspects of progressive thought, whether they acknowledge this influence or not. Further, tendencies to downplay or even denigrate working-class culture are not merely artifacts of the past. As scholars in other fields have begun to point out, within the middle-class dominated environments of universities progressive ideas about democratic and deliberative practice still broadly pervade thinking about democracy across academic disciplines.[5]

The field of education also provides a useful case study for other fields because education has always been seen by progressives as one of the most critical arenas (perhaps *the* most critical) for interventions to foster a more democratic society. It was no accident that Dewey started a school, even if he later lost faith in schooling as an independent avenue for social change. And his vision of social change remained "educational" to the end. As recent scholars like Fred Rose and Paul Lichterman have shown, middle-class progressive activists still hold tight to a deep faith in education and individual change as the key fulcrum of social change today.[6]

As a counterweight to progressive visions of democracy, in Chapter 6 I lay out a working-class alternative that I call "democratic solidarity." Versions of this model have long been prevalent in a range of working-class-dominated settings, especially labor unions. I look in particular to what is generally called the field of "community organizing" in the tradition of Saul Alinsky as a key example of how "solidarity" can be made "democratic" in ways classic progressives have seemed unable to recognize. Organizers like Alinsky have sought to confront inequality directly with mass mobilizations instead of trying to slowly shift the broader culture toward what they have generally seen as progressives' unreachable, utopian models of collaboration, egalitarian exchange, and reasoned negotiation. Proponents of democratic solidarity seek to make the empowerment of those at the bottom rungs of our material and social world a realistic possibility in the here and now. Alinsky's writings provide an example of the ways working-class organic intellectuals have reacted against middle-class efforts to enforce what they see as progressives' privileged fantasies. From the perspectives of Alinsky and others, progressive exhortations to "wait" embody a reprehensible paternalism on the part of those who do not really understand what it is like to *suffer*.

I am deeply sympathetic to the working-class vision of empowerment and disturbed by its absence in the educational literature and elsewhere. But I do not argue that working-class forms of democratic solidarity should simply *replace* visions of progressive democracy. Instead, I examine the contrasting strengths and weaknesses of each conception. In the best of all possible worlds, efforts to foster democratic empowerment would draw from aspects of both progressive and working-class strategies.

Such a synthesis has proved extremely difficult to achieve, however. In part this is because cultural groups on both sides have generally failed to see what is worthy in the action practices of others. This volume is meant as a contribution to a broader effort to challenge these cultural blindnesses. Efforts to integrate different approaches, however, are also complicated by inevitable inequalities of power (the very inequalities that progressives have often downplayed). When middle-class professionals come into settings previously controlled by members of the working-class, for example, they often end up dominating, unconsciously enforcing their own cultural ways of speaking and acting, leading to the departure of those less equipped to participate in this manner.[7] This volume does not attempt to solve this problem, although I have begun to explore this issue in other related writings.

The penultimate chapter of this book provides a case study of how different approaches to democracy and empowerment played out in the real world during the civil rights movement in the South. The case study also shows how the clarity of the relatively abstract visions discussed in previous chapters becomes complicated and how these visions often interweave with each other in unexpected ways in the contingency of actual social contexts. And it contests the (usually implicit) tendency of education scholars to justify their use of progressive pedagogies for student empowerment by pointing to the civil rights movement as a clear example of progressive democratic organizing.

This volume concludes with a discussion of the implications of these findings for schools and scholarship on democratic empowerment more broadly. At the same time, I speculate on the kinds of useful roles middle-class academics may play in bringing non-middle-class visions more centrally into the academy.

Some of the chapters that follow incorporate versions of articles published previously elsewhere. Chapter 1 is based on "Social Class and Social Action: The Middle-Class Bias of Democratic Theory in Education" and Chapter 2 on "John Dewey's Conundrum: Can Democratic Schools Empower?" published in 2008 and 2001, respectively, in *Teachers College Record*. Chapter 3 is based on "John Dewey and 'a Paradox of Size': Faith at the Limits of Experience," published in 2001 in *American Journal of Education*.[8] Those who want a somewhat more detailed discussion of the issues addressed in Chapters 2 and 3 might benefit from a look at the original articles. Sections of some of these articles also appear in other chapters where relevant. These articles were written at different times, and I did not attempt to bring them fully up-to-date with the most recent literature except where this seemed critical. I have also changed some of the terms I use here from those used in the articles. For example, in "Social Class and Social Action" I referred to what I now call the "collaborative" progressives as the "democratic" progressives. Since the personalist group is also democratic in its own way, I increasingly saw that the earlier phraseology would have been confusing here.

Part I
Overview

1
Social Class and Social Action

> Progressives ... intended nothing less than to transform other Americans, to remake the nation's feuding, polyglot population in their own middle-class image.
>
> —Michael McGerr, *A Fierce Discontent*

> From the beginning the American intellectual ... [chose] a paradoxical vocation: a social critic committed at once to identification with the whole of the people, and an elitist whose own mores and life situation proved somewhat alienating from the very public he or she had chosen to serve.
>
> —Leon Fink, *Progressive Intellectuals and the Dilemmas of Democratic Commitment*

Introduction

At the end of the 1800s, American intellectuals began a long if occasionally interrupted romance with progressive visions of democracy. For more than a century since then, scholars across the social sciences and humanities have found different aspects of progressive democratic practice extremely compelling, even though few if any of their hopes for social transformation have ever come to fruition.

Why?

A core motivating factor, I argue, has been social class. From its earliest beginnings, progressivism, writ broadly, reflected the desires and beliefs of middle-class professionals in America. As a result, the democratic models embraced by progressives embodied, in different ways, the cultural patterns and preferences of the middle-class intellectuals who developed them.

This chapter provides an overview of the broad argument of this book. It begins by tracing the emergence of distinct middle and working classes in America at the end of the nineteenth century, showing how progressivism emerged as an integral part of this process. I describe how three distinct branches of progressivism emerged, which I call "administrative," "collaborative," and "personalist," developing out of a shared set of social concerns and cultural practices. Together, these conceptual perspectives provided the middle class with ways to explain each branch's distinctive "truths." Progressives used these social frameworks to map out borders between themselves and "others," distinguishing between cultural groups that were more and less prepared to adequately perform the duties of modern citizens. Those not from mainstream middle-class backgrounds, not surprisingly, did not fare well in this analysis.

At the same time, the middle and working classes became increasingly distinct; their ability to understand and relate to each other diminished. Contrasting forms of what I call "democratic solidarity" predominated in working-class settings. This was especially evident in labor struggles. I focus, here, on the model of "community organizing" developed by Saul Alinsky and organizers who came after him. I show how community organizing maintained a deep commitment to democracy even though it gave less emphasis to the individual creativity and expressiveness prized by progressives. Community organizers pragmatically stressed the importance of enforcing a collective "voice" in public to gain power in the here and now.

My point in this book is not to deny the sophisticated insights of progressive thought. In fact, I explore many of these in the chapters that follow. Instead, I seek to place progressive ideas within a larger spectrum of possible ways of being "democratic," balancing middle-class commitments and concerns with those of a working class facing very different material and social challenges. And regardless of their sophistication, progressive democratic dreams will not serve us well until we acknowledge the implicit, and too often explicit, classism (and associated racism) that has come with these dreams.[1]

This book focuses on the context of education. The educational visions of the progressives provide an especially useful case study of the development of democratic practice, in part because progressives themselves always focused on education as a key site for social action and change. In fact, as we will see, it was in John Dewey's Laboratory School, in the progressive schools of the 1920s, and in the free schools movement of the 1960s that progressive activists and intellectuals created some of their most fully fleshed-out examples of the forms they hoped a broader progressive society might embody.

Democracy and Education

Over the past few decades in American schools, progressive visions of democratic education have largely fallen away. Especially in the last ten years or so, in the wake of No Child Left Behind, conversations about education have increasingly focused on narrow conceptions of learning. Visions of a better society and of more fulfilled human beings have given way to a stress on efforts to improve students' job prospects and the larger economy.

Of course, more idealistic visions of education have always been honored more in the breach than in reality. Schools have always been places where children mostly learn to replicate the class positions of their parents.[2] Nonetheless, there have been moments in the past where groups of progressive educators and scholars not only embraced more expansive visions of education but also found ways to insert these ideas, however marginally, into classrooms, new schools, and the curriculum. In fact, until quite recently educating children for democratic citizenship was a core value in Americans' views of the goals of schooling. As late as the 1960s, Americans still saw schools as key pillars of a democratic society—regardless of how vaguely or problematically this may have been framed.[3]

While scholarship in education reflects to some extent the narrowing of the curriculum we see in actual schools, broad holistic visions of education have remained compelling to many "progressive" educators and scholars. The popularity of Nel Noddings's vision of caring classrooms that nurture the unique individuality of students, as well as the dominance of John Dewey's vision of democratic education—even when Dewey himself is not explicitly referred to—are both good indicators of this.[4]

At its core, then, the field of education is still driven by dreams of an egalitarian society. Progressive scholars still hope that teachers might, at least sometimes, reach beyond the façade of formal schooling to fan the flames of the unique capacities of individual students. In fact, in contemporary schools of education, where the vast majority of educators are trained, David Labaree has found "a rhetorical commitment to progressivism that is so wide that, within these institutions, it is largely beyond challenge." Educational scholars, then, remain intellectually and emotionally committed to a conception of "the school as a model democratic community" and to "making the reform of education a means for the reform of society as a whole around principles of social justice and democratic equality."[5]

Theorizing about Social Class

From the quartet of theorists who have most influenced our views of class in the Western intellectual tradition—Karl Marx, Max Weber, Emile Durkheim, and Pierre Bourdieu[6]—my analysis is informed primarily by Bourdieu. The first three tended to focus on interrelationships between class and the economic structures of capitalist society. While many of their basic assumptions form the background of my story, my central interest is in the sociocultural effects of these economic developments. For these purposes, Bourdieu's work seems most relevant.

Most important, for my analysis, is Bourdieu's conception of "cultural capital." Bourdieu argued that social practices in society represent a form of capital different from, and yet in some cases as important as, economic capital.[7] Capitalist society is stratified, then, not only in terms of the "material" resources of different groups but also in the relative value of the different cultural practices that these groups tend to embody.

His conception of the relationship between what he called "habitus" and "field" provides the foundation for his vision of cultural capital. A habitus is the set of social practices and dispositions associated with a particular social position. One way to think of a habitus is as a bundle of interrelated strategies for responding to a group's "conditions of existence."[8] And every habitus is designed to respond to a particular social "field." For example, a person with a middle-class habitus at the turn of the twentieth century would have had little understanding of how to act appropriately in a working-class saloon, whereas a manual worker would feel just as lost in a lawyer's office.[9]

Informed by Bourdieu's general ideas about culture, this chapter maps out key characteristics of middle- and working-class culture as they emerged in the United States. In contrast with Bourdieu's rich, multifaceted models of class structures,[10] and unlike many other scholars working on the structure of class in postmodern or postindustrial societies, I focus on two positions—the middle and working classes.[11]

Because middle- and working-class cultures exist nowhere in the world in any "pure" form, I employ these terms as what Weber called "ideal types."[12] As Alvin Gouldner argued, "clarity" in social analysis "is always dependent not on good, but on *poor* vision; on blurring complex details in order to sight the main structure."[13] Scholars synthesize different ideal types in response to particular questions. If one is interested in the distribution of different kinds of "occupations" in a society, for example, one may end up with a large number of "classes."[14] For the purposes of this analysis, the binary formulation has seemed most productive, reflecting what emerged through my examination of the evidence as two relatively coherent historical strands of practices (habituses) and social contexts (fields).[15] I refer only in vague terms to a third group, the upper class of society that owns and in some cases directs the institutions in which the middle and working classes labor and live. This vagueness is, in part, a product of an increasing complex system of capitalist control that makes it difficult to identify "who" is in control.[16]

Today, only a limited segment of society seems to embody middle- and working-class traditions in any substantial sense. What I am calling middle-class cultural patterns remain most prominent among members of the "upper" middle class: managers, analysts, and professionals who retain significant independent power within and outside the corporate entities that rule much of our economic life.[17] Working-class traditions, in contrast, seem most evident today in the daily practices of labor unions and among workers who remain deeply rooted in long-term relationships with local communities and extended families.[18]

Social Class in the United States: A Brief History

To understand the traditions of social class in America, it is important to have a sense of the historical trends and social and material conditions that helped produce them. I begin with a brief summary of the history of the emergence of the middle and working classes in America, and then discuss how these early cultural trends in some cases intensified and in other cases fragmented and blurred during the twentieth century.

The Emergence of the Middle Class

A substantial middle class did not emerge in America until the middle decades of the nineteenth century.[19] Before that time, there were what Stuart Blumin called "middling" folk: small farmers, skilled workers or artisans, shopkeepers, and the like. These "middling" folks were of modest means compared to the elite citizens of their day, their relatively low social status deriving not only from their limited income but also from the fact that they generally engaged in manual labor. By the middle of the nineteenth century, however, pressures of industrialization had begun, slowly, to dissolve this "middling" group. As firms grew larger and more complex, local manufactories and home-based businesses were replaced by companies and corporations.[20]

Firms began to separate manual laborers from "clerks" and other non-manual workers who handled paperwork and sales, among other duties. First, in small concerns, they simply worked in separate rooms. But as cities became more spatially specialized they increasingly worked in completely separate locations. Over time, this distinction between manual and nonmanual labor became the key indicator of nineteenth-century class status. By the 1890s, manual and nonmanual workers increasingly inhabited "separate social world[s]" as cities became segregated by class.[21]

The increasing complexity of the world created by industrialization that accompanied the transition from "middling" to middle class was very confusing for the members of this evolving group. They had to develop new ways to keep their footing in the shifting sands of modernity. Rapid urbanization fragmented personal networks, as the ability to transfer "status from one place to another . . . eroded." In an increasingly anonymous world, the old systems of patronage and letters of introduction lost their controlling force. In response to the loss of tightly woven networks of personal relationships, the middle class developed more objective standards and qualifications for particular jobs that allowed people to act as relatively autonomous individuals. "Diplomas and degrees, accreditation boards, registrars, government identification papers, licenses, and later more standardized impersonal testing helped individuals and groups navigate through and deal with anonymity." At the same time, the middle class developed a diversity of associations that "evolved a range of organizational procedures to deal with their increasing size and impersonality."[22]

These changes required the development of a broad new set of social practices and self-understandings that could allow the members of the middle class to successfully orient themselves in this new "impersonal" world. "One had to forge a self-reliant, confident, and independent sense of identity cut free from reliance on the approbation, support, or referencing of friends, for such contacts were short-lived and less reliable through time." There was increasing criticism of "cronyism," although this did not, of course, disappear. "Privacy, confidentiality, and nonjudgmental impartiality, rather than acting for one's 'friends' . . . gradually emerged as the new ethical ethos of the middle-class life." Through these efforts to forge a more independent, objectively defined identity "would emerge the more modern sense of self that defined the new middle class."[23]

The increasing wealth of the middle class allowed them to purchase larger residences separated from the homes of the "masses," with multiple rooms for different activities. In these new contexts a middle-class "domestic" ideal began to emerge, altering gender roles and "strategies of child nurturance and education." The new middle class "'initiated methods of socialization designed to inculcate values and traits of character deemed essential to middle-class achievement and respectability,' values and traits not of the aggressive entrepreneur but of the 'cautious, prudent small-business man.'"[24]

At the same time, partly in order to concentrate their resources, middle-class families began to limit their size. Within the frame of the new domestic ideal, the experience of children in these homes was transformed. Perhaps most important, "children were given greater amounts of formal schooling, a crucial tactic intended to help them secure positions in the expanding nonmanual work force."[25] In fact, for a range of reasons, as we will see, a college degree quickly became a key indicator of middle-class status.[26]

As a result of interactions between the changing conditions of their lives and the social strategies they developed in response, members of the middle class increasingly defined themselves by their abstract "qualifications" and by their separation from the dirty experience of manual work. Their world increasingly became dominated by numbers and file cards and identifiable formal knowledge. Because "no abstract representation on paper . . . conferred the knowledge that sight and touch did," middle-class workers became "lost" in "numbers, forms, charts and rules," becoming relatively "bodyless" in contrast with the emphatically embodied existence of the working class.[27] At the same time, a sober, "Victorian" vision of life and duty began to emerge among the middle class.

During these decades the middle class became an odd kind of "class" that maintained a coherent collective identity through a kind of studied independence. As Blumin noted, this "brings us face-to-face with a central paradox in the concept of middle-class formation, the building of a class that binds itself together as a social group in part through the common embrace of an ideology of social atomism." A "new character ideal" emerged in this impersonal world: "the team player" able to continually shift relational ties and work closely with relative strangers.[28]

The Emergence of the Working Class

Woe unto the man who stood alone in this pitiless struggle for existence.

—David Montgomery, *The Fall of the House of Labor*

Similar processes of industrialization also molded a new working class. At the beginning of the nineteenth century, an enormous class of wage laborers had been almost unthinkable. But by the end of the century, "wage labor emerged . . . as the definitive working-class experience."[29]

The conditions of industrial work, which by 1900 had captured "more than one third of the population," differed in fundamental ways from those of "white-collar workers." Middle-class, nonmanual workers maintained significant independence, increasingly depending on individual expertise for their continued success. In contrast, in factories the holistic skills of artisans were systematically broken down into separate operations, allowing the hiring of much less skilled workers, holding wages down, and threatening workers' independence on the worksite. By 1886, 65 to 75 percent of the labor force was semi- or unskilled. Furthermore, in contrast with the clean offices of the nonmanual class, working-class labor "was often dirty, backbreaking, and frustrating."[30] Factory workers at the end of the nineteenth century increasingly worked under the "clock," laboring in settings ruled by "compulsion, force, and fear."[31]

The uncertain existence of manual workers was made even more difficult by the fragility and unpredictability of the nineteenth-century economy. The nation stumbled from depression to depression. In 1875, for example, only one-fifth of the population could find regular work.[32] During the 1880s and 1890s, business failures rose as high as 95 percent.[33] As has always been the case, those on the bottom suffered the most through these tumultuous times, as wages in real terms for manual workers fell.[34] By the end of the 1880s, "about 45 percent of the industrial workers barely held on above the $500-per-year poverty line" and "about 40 percent lived below the line of tolerable existence."[35] In fact, "inter-class mobility disappeared" for most as early as the 1850s, as "the membership of the classes became" increasingly "fixed."[36]

As wage labor became an increasingly central part of modern life, workers responded with expressions of solidarity, seeking to contest the predations of the industrial age. Workers fought in the industrial realm for wages and other concessions, as well as in the political realm for legislation mandating reduced work hours among many other issues, focusing at different times on one or the other avenue. In the first half of the nineteenth century, the labor organizations that formed during good economic times were repeatedly destroyed in the myriad depressions. By the Civil War new organizations increasingly realized they needed to create structures and develop resources that would allow at least some groups to survive through the bad times. But despite some important successes—especially in legislation—and thousands of strikes, peaking and falling with the waxing and waning of prosperous times, labor still mainly faced defeat.

At different moments, an incipient working-class consciousness seemed to be emerging. Although the great railroad strike of 1877, like many others, was brutally put down by state and federal forces, for example, sympathy strikes spread through many communities, and a broad mass of working-class citizens supported the strikers. In fact, some militias sent to suppress the strikers ended up joining them instead.[37] But a sense of common cause did not ultimately coalesce in America. Manual workers remained fractured by racism, sexism, and a range of ethnic, religious, urban and rural, immigrant and "native," and skilled craftsmen and unskilled laborer conflicts. In fact, one of the most common strategies for self-defense involved attempts to exclude "others" from employment. The "mutualism" of working-class life could just as easily feed group division as collective solidarity.[38]

Despite these internal differences, class distinctions between workers and the more privileged classes became increasingly evident, especially in the burgeoning cities. Members of different classes easily recognized each other as what they were—by the way they walked, the way they talked, the clothes they wore, and so on. "Some workers, by no means all," since these developments were always uneven, came "to occupy a separate social world within the antebellum [post–Civil War] city—their social networks can be reasonably described as consisting almost entirely of other workers."[39]

As the middle class developed its culture of domesticity, individualism, and restrained association, the working class necessarily depended upon very different forms of collective solidarity—of families, of communities, of trades, and more. In crowded neighborhoods, "the constraints and uncertainties of working-class life—low wages, lay-offs, accidents, limited opportunity, early death—made individualism at best a wasteful indulgence and at worst a mortal threat." Under these conditions, workers developed "a culture of mutualism and reciprocity," teaching "at home and work . . . sometimes harsh lessons about the necessity of self-denial and collective action." In fact, "daily experiences and visible social distinctions taught many workers that although others might wield social influence as individuals, workers' only hope of securing what they wanted in life was through concerted action."[40] While the middle class increasingly lived in a world of acquaintances and strangers, then, the working class depended on how embedded they were in long-term ties.

In the "cramped living spaces" of the working class, "in slum tenements or abandoned middle-class housing in older districts," the domestic ideal aspired to by the middle class was largely unreachable.[41] Lacking substantial opportunities for individual or family privacy, working-class residents participated in "shifting communities of cooperation [that] had none," or at least substantially fewer, "of the counterbalancing elements of the female domestic sphere of calm and affection that bourgeois men and women prized."[42] Poverty meant that everyone generally had to work. And these facts of life had important implications for childhood in these settings. The "conditions" of working-class life "made it that much harder" for working-class children "to develop a sense of individuality and autonomy"[43] that was so celebrated by middle-class families. In fact, efforts to assert middle-class forms of autonomy were often seen as threatening to the survival of the family unit as well as at work and in the extended relational ties of working-class communities.

Shifting Forms of Social Class in the Twentieth Century

i sit here all day and type
the same type of things all day long ...
day after day/adrift in the river of forms ...
i am a medical billing clerk
i am a clerk.
i clerk.

—Wanda Coleman, "Drone"

The twentieth century brought vast changes in the structure of the national and global economy and increasingly complex, overlapping layers of social diversity. For the working class, the most important shift, as Harry Braverman noted, was probably the growth of a broad range of non-middle-class service jobs whose work embodied many characteristics of working-class labor but looked very different from manual labor in factories and elsewhere.[44] Initially most visible as a vast increase in low-level office workers (mostly women), an enormous army of low-pay positions emerged in sales, food service, hospitals, janitorial services, and more recently, call centers.[45] Braverman argued that these new positions were clearly working class, subjected to the process of "deskilling" familiar to earlier manual workers.[46] Nonetheless, the recent explosion of new kinds of positions with a range of different job requirements (e.g., technicians and a complex proliferation of health care jobs) has clearly complicated and blurred any simple binary distinction between middle and working classes.

Throughout the twentieth century, fairly strict hierarchical control has remained much more evident at the lower levels of firms than at the top. And capacities for control have been magnified by new systems of "scientific management" instituted after the turn of the century, intensified recently by sophisticated information technologies. In recent years there have been some efforts around (or at least rhetoric about) providing opportunities for more individual discretion and encouraging more collaboration among nonmanagement workers. While some scholars question whether these efforts have substantially altered the work environment of low-level employees,[47] this new focus on encouraging teamwork at all levels of a firm may also contribute to a progressive blurring of clear distinctions between middle- and working-class jobs and discursive practices.

While the experience of work among lower-level employees has fragmented to some extent, evidence indicates that the importance of middle-class practices of teamwork for managers and professionals has only increased. As David Brown argues, because these workers are relatively autonomous, organizations cannot set strict guidelines and are forced to depend on social "norms ... that facilitate control from a distance ... together with structural policing mechanisms such as committee work (where 'colleagues' police one another)."[48] As the "postmodern" workplace advances it seems likely that these pressures for self-guided collaboration at the higher levels will continue to intensify.[49]

Outside the realm of work, a range of social and material changes in our increasingly postindustrial world has also complicated the structure of social class in America. For example, the strong local working-class communities that provided an important grounding for earlier working-class cultures have largely disappeared in many areas. This loss of community is especially evident in the impoverished, segregated areas of our cities.[50]

For managers and professionals, in contrast, the growing fluidity of postmodern life and their progressive loss of connections to particular places and communities seem, for most, to have largely magnified cultural trends already visible at the end of the nineteenth century.

Key Characteristics of Middle- and Working-Class Cultures in America

Patterns of Middle-Class Life

A wide range of studies have shown that the standard parenting practices of the middle class today are significantly different from those of working-class families. Middle-class children learn at an early age to monitor themselves and make their own judgments about the world. In fact, these children are often encouraged to participate in adult life as if they were "mini" adults themselves. They are frequently asked for their opinions and are allowed (and even encouraged) to express disagreements about adult directives.[51] Middle-class parents celebrate children's unique characteristics and capabilities, helping them develop a sense of themselves as discrete and unique individuals. As a result, their children often begin to feel an "emerging sense of entitlement."[52]

Even as middle-class families promote independent thought, however, their discourse patterns tend to make "the insides of [family] ... members ... public,"[53] providing a powerful tool for closely monitoring individuals' thoughts and ideas. This continual monitoring makes it possible for middle-class parents to nurture the development of "internal standards of control" and allows them to downplay the need for strict rules and guidelines for children.[54] The spatial privacy often made possible by the size of middle-class residences, then, is joined with an often extreme lack of psychic privacy.

In their discursive interactions with children and each other, middle-class parents tend to prefer forms of relatively abstract reasoning. Echoing other studies, Betty Hart and Todd Risley found, for example, that professional parents "seemed to be preparing their children to participate in a culture concerned with symbols and analytic problem solving."[55] And many have noted that these discourse patterns fit well with the kind of institutional and employment situations that these children will participate in throughout their lives.[56] In our increasingly information-driven world, middle-class managers, symbolic analysts, and other professionals increasingly focus on the manipulation of relatively abstract data. Even when middle-class workers engage more directly with the contingencies of the real world—think of surgeons or engineers—their work is generally deeply embedded in a broad milieu of abstract data and symbolic relations.

The lives of middle-class children are also highly structured and scheduled, leading them to spend much less time than less privileged children on informal activities and child-directed play. In fact, middle-class parents focus so intently on their efforts to "cultivate" their children that their "lives" can have "a hectic, at times frenetic, pace of life."[57]

The frenetic existence of middle-class childhood, with its shifting cast of characters, fosters mainly "weak" social ties. Children learn to interact with a wide variety of relative strangers and are less likely to be embedded in tight networks of extended family relationships.[58] This tendency is magnified by the isolation of nuclear families and the relatively high mobility of middle-class people, who frequently leave home for college or employment and never return.[59] Despite the weakness of their ties and their lack of rootedness in local communities, the connections made by the middle class generally give them access to more resources than the less privileged. Because they share the discursive and cultural practices of other privileged people, they can interact with them as relative equals.[60]

Finally, collaboration and teamwork have become increasingly central characteristics of middle-class life over the twentieth century. Group success often requires managers and professionals to work closely with people with whom they have no long-term relationship. Each individual in these contexts is expected to independently contribute his or her own particular knowledge and skills to an often weakly defined common project. Collaboration in these groups is facilitated by the relatively abstract, elaborated discourse predominant in middle-class settings.[61] I refer to this particularly middle-class form of joint action as *collaborative association*.

In fact, a broad range of research has indicated that the key characteristic of middle-class employees is not any specific knowledge they may hold but their internalization of the general practices of middle-class discourse and interaction. Because these workers are relatively autonomous, organizations must be able to trust that they will independently support the goals of the firm. Under conditions where they must engage with a broad and unpredictable number of relative strangers, white-collar workers focus their energies on maintaining "standardized and routinely sanctioned patterns of behavior."[62] In her interviews with upper-middle-class men, for example, Michele Lamont found that "for American professionals and managers, the legitimate personality type rewarded by large organizations presents the following traits: conflict avoidance, team orientation, flexibility, and being humble and not self-assuming."[63] Because professionals face situations that generally lack clear guidelines, involve the manipulation of data, and require frequent interaction with relative strangers, they focus their energies on maintaining the "standardized and routinely sanctioned patterns of behavior" that mark them as middle class in multiple contexts. [64]

Lamont also found that, given the shifting goals and guidelines they encounter, for upper-middle-class men "living up to one's moral standards is often constrained by situational factors ... often conflict[ing] with pressure for conflict avoidance and team orientation." In fact, "to a certain extent the cultural imperative for flexibility prevents ... [them] from putting personal integrity ... at the forefront. Indeed, some might end up adopting a pragmatic approach to morality as they adapt their beliefs to the situation at hand."[65]

As Brown noted, these tendencies help explain the requirement of most middle-class jobs for a college degree of some kind, often with little attention paid to the content of what was studied. Because middle-class people are more likely to operate within settings with less stringent controls over their action, organizations are forced to depend on "norms ... that facilitate control from a distance ('responsible' behavior and 'disinterestedness') together with structural policing mechanisms such as committee work (where 'colleagues' police one another)"—in other words, on how *middle class* these employees are. In college, students learn a "fairly standardized type of language or 'code'" that will serve them well in these settings. As a result, college produces "a relatively uniform character type" that can be "expected to get along with other employees, especially fellow graduates."[66] In fact, there is a reciprocal relationship between higher education and middle-class status, then. Arriving at college fluent in middle-class practices makes success more likely, and success progressively strengthens one's cultural identification with the middle class.[67]

Higher educational institutions are central places for nurturing middle-class dispositions. This is part of the reason that the paradigmatic experience of upper-middle-class late adolescents is leaving home to attend a residential college with an established reputation. The structures of the laboratory, the seminar, and even the didactic lecture embody the abstract, dialogic practices of middle-class managers and professionals.[68] Professors and students at four-year institutions live in a social world dominated by middle-class values and practices, a world that actively excludes and marginalizes manifestations of working-class ways of being but that rarely acknowledges this exclusion. And as students move through higher and higher levels of education, success requires ever more fluency in middle-class forms of discourse and interaction. At the highest levels, in doctoral programs, only middle-class ways of framing problems and issues or of presenting the results of research are generally legitimate.[69]

For the middle class, there is a clear continuity between these different aspects of their lives. Children and their parents move relatively easily between home and school and work. They encounter others who they interact with on a relatively equal level and who think and act much like they do. In all these contexts their facility with abstract knowledge, their sense of individual entitlement, and their skills at discursive social interaction serve them well. It should not be surprising, then, that the work of many middle-class adults is often tightly integrated into their private lives. They tend to have "careers" rather than just "jobs." As Lamont noted, "in contrast to blue-collar workers," the upper-middle-class men she interviewed "rarely live for 'after work.'"[70]

Patterns of Working-Class Life

> Overtime is a delicacy gobbled
> by family men who wipe their mouths
> and say Baby needs new shoes.
>
> —Todd Jailer, "Chester Gleason"

Annette Lareau found that "in working-class and poor homes, most parents did not focus on developing their children's opinions, judgments, and observations."[71] Instead, their families were structured to a much greater extent around an established hierarchy between children and adults. Some have argued that these patterns are partly a result of the hierarchical conditions of working-class labor.[72] More pragmatically, because working-class parents lack time to constantly monitor children, hierarchies and limited tolerance for "back talk" make more sense than constant negotiation.

Although working-class parents seem less focused on encouraging individual expression, working-class children often have more frequent opportunities for child-initiated play than children in middle-class families. In contrast with what she termed the "concerted cultivation" approach of the middle class, then, Lareau argued that working-class parents are more likely to "engage in the *accomplishment of natural growth*, providing the conditions under which the children can grow but leaving leisure activities to children themselves."[73] These relatively open contexts for play provide alternate avenues for individual expression, including forms of dramatic storytelling that express both individuality and the ways that individuals are embedded in long-term relational ties with others.[74] Access to an audience is not simply given to children in working-class settings, however. In such contexts, Peggy Miller, Grace Cho, and Jenna Bracey found, "working-class children had to work hard to get their views across; . . . [they] had to earn and defend the right to express their own views."[75] There is little entitlement here.

"Working-class people" in the United States "are more likely to live where they grew up, or to have moved as a family and not solo. They are more likely to live near extended family and [are] . . . likely to have been raised and socialized by traditionally rooted people."[76] Even though the old ethnic enclaves of the nineteenth and early twentieth century have largely disappeared, Alfred Lubrano found that a "core value of the working class" still involves "being part of a like-minded group—a family, a union, or a community."[77] As at the end of the nineteenth century, today this tendency to value deep connections with families and communities is partly driven by the material conditions of working-class life. Many workers have no choice but to depend on a web of links with others to get them through hard times, and, as I have noted, the impoverished, especially in the central cities, suffer greatly to the extent that these relationships have fractured or lack significant resources. In a world of globally increasing inequality, Zygmunt Bauman has stressed, those on the bottom "are 'doomed to stay local,'" where "their battle for survival and a decent place in the world" must be "launched, waged, won, or lost."[78]

Some have argued that working-class labor is relatively simple compared with that of the middle class,[79] but the evidence indicates that this issue is more complex. Although employers have sought for more than a century to reduce workers' discretion and skill, a range of studies have shown that many seemingly basic fast food, data entry, industrial, and other working-class jobs actually require extensive learned capacities.[80] In fact, Trutz von Trotha and Richard Brown argued that the strict guidelines characteristic of many working-class jobs, which cannot hope to capture the subtlety of actual work, actually end up forcing workers to "incessantly focus on the cues and clues of specific situations to discern, or invent ad hoc, the meanings and actions that might be appropriate." "Generally speaking," they concluded, "the lower class person considers a wider range of imponderables, and can take less for granted, than does the middle-class actor."[81] In other words, while managers and professionals may face a higher cognitive load in realms of relative abstraction, workers are more likely to face more (but equally complex) concrete challenges in their local environment. A key tendency of working-class labor, therefore, is not its relative simplicity but instead its relatively embodied and tacit nature.[82] Even when extensive abstract thought is required (for a carpenter, for example), this is likely to be deeply embedded in material requirements of a specific job.

To middle-class managers, different devices have "parameters," but for workers, individual machines can actually have different "personalities." This tacit and embodied character of working-class experience partly explains why one can usefully include highly skilled craft workers and low-skill line workers, who can be trained in twenty minutes, in the same "class." As Fred Rose noted, "the working-class experience of physical labor teaches people to trust the practical knowledge gained from personal experiences" over the generalized knowledge of research.[83] Basil Bernstein similarly distinguished between a working-class tendency to "draw upon metaphor," and a middle-class focus on abstract "rationality."[84]

The truth is that employers are at least as dependent upon the innovations of working-class people as they are on those of middle-class employees. But while the innovation of the middle class is often explicitly and actively encouraged and rewarded, the ongoing innovation of the working-class tends to progress invisibly below the level of employer dictates. In fact, working-class innovation actually operates counterintuitively as a kind of *resistance* to the strictures of the system, even though this "resistance" is actually what allows the system that oversees them to continue.[85] The same thing can be said of middle- and working-class processes of learning. While the middle class is often rewarded for acquiring knowledge, the "informal learning" on the job and in families and communities that "has been heavily relied upon to actually run paid workplaces" and that dominates working-class community life remains largely "unrecognized" by both employers and educators. Thus, firms "appropriate ... the production knowledge of workers without valorizing or compensating it."[86]

One result of the different forms of knowledge celebrated by the middle and working classes is that each, for different reasons, often sees members of the other class as relatively "stupid." Thomas Gorman found, for example, that "members of the working class hold an image of the middle class as being incompetent in negotiating everyday events and having knowledge that is not practical."[87] In the extreme, as Lubrano noted, the middle class can be seen as the kind of people who have to hire someone to change a light bulb. And this ignorance of the middle class sometimes empowers workers. Susan Benson, for example, described how working-class saleswomen in early department stores maintained control over their work in part because their managers found it distasteful and challenging to descend into the messy complexity of the actual selling process from the familiar abstractions of their office paperwork.[88] At the same time, through countless "injuries" experienced in their interactions with the middle class, members of the working class are very conscious of the fact that the middle class tends to look down on them.[89] And in Gorman's study "one half of the middle- . . . class respondents" did, in fact, make "blatantly negative comments towards members of the other social class."[90]

Given the contrasting conditions of their lives, the working class has developed different practices of interpersonal engagement and strategies for orienting group activity. On the most basic level, workers tend to prefer a different set of values in their co-workers and friends than members of the middle class. Relatively flexible middle-class attitudes about morality and reverence for unique individuality contrast strongly with working-class tendencies to stress the importance of tradition, personal integrity, personal responsibility, sincerity above flexibility, and the quality of interpersonal relationships.[91] They are more likely to prefer "straight talk" and "resolving conflicts head on," as opposed to placation and long discussions.[92]

Operating in situations where embodied knowledge dominates and where coordination requires mutual adjustment amid an ongoing flow of work, the working class depends less on collaborative association than on what I will call *organic solidarity*.[93] In contrast with the focus on individuality characteristic of middle-class settings, working-class groups are more likely to operate as a collective unit.

It is important to emphasize, however, that these rich "communalized roles" are "strikingly inconsistent with a picture of lower-class" groups' work as relatively simplistic forms of "mechanical solidarity."[94] In important ways, organic solidarity is itself a form of collaboration that can be as responsive to individual capacities and interests as the more explicit forms of collaborative association preferred by the middle class. Lacking time for extensive negotiation and dialogue, it should not be surprising that this approach to joint action is generally grounded in established, if sometimes informal, hierarchies.

Although lower-level workers often seem invisible to the relatively privileged, the working class continually deals with the power of managers and professionals to affect their lives in profound ways.[95] In fact, in their interactions with middle-class institutions beyond their private spheres—especially in schools and work sites—working-class people often feel relatively powerless.[96] They often resent "middle-class language . . . and middle-class attitudes."[97] Yet those on the lower rungs of America's economic ladder often also feel extremely dependent on the middle class, especially for the advancement of their children.

While middle-class parents know, instinctively, how to prepare their children to succeed in middle-class settings, working-class parents often do not. With respect to schooling, for example, they often "believe that they can be most helpful by turning over responsibility for education to educators." At the same time, however, Lareau and Wesley Shumar found "in interviews and observations, [that] working-class and lower-class mothers repeatedly expressed fear that the school would turn them in to welfare agencies and 'take their kids away.'"[98]

Making the situation even more difficult, we know that working-class children tend to get a "working-class" education in schools. The experiences of many of these children in classrooms, then, are unlikely to provide opportunities to learn middle-class practices and forms of discourse.[99] Ironically, middle-class children are more likely to succeed even in school settings framed by working-class culture. They are much better equipped to adjust to the forms of abstract knowledge and discourse demanded by even the most didactic classroom. And because middle-class children are initiated into middle-class practices before they get to school,[100] it matters much less for them whether teachers provide them with more engaged and interactive middle-class experiences. In other words, those who may "need" initiation into middle-class practices not only don't get them but also couldn't easily appropriate them even if they did get them, while those who get middle-class practices in schools often don't really need them.[101]

The tensions between middle-class and working-class ways of being can become especially intense when working-class people go to college. College can involve "a massive shift . . . requiring an internal and external 'makeover.'"[102] In fact, Peter Kaufman's study found that the most successful working-class college students were those who were most able to disassociate themselves from their old friends and their old community.[103] Helen Lucey, June Melody, and Valerie Walkerdine similarly found in their interviews with working-class women that "wanting something different, something more than your parents, not only implies that there is something wrong with your parents' life, but that there is something wrong with *them*." Successfully entering the middle class often requires working-class people to embody a "split and fragmented subjectivity" that can allow them "to cross the divide."[104] Such bicultural fluency is difficult to achieve and sustain, however.[105] Completing a residential four-year college degree away from home, then, is both the best way to become middle class and one of the most powerful ways to alienate oneself from one's home community.

And working-class parents can be less than supportive or understanding of college dreams. In fact, given the "hidden injuries of class" they often experience, it turns out that "having middle-class contacts ... not only does not guarantee that the working class will raise their educational aspirations," it can have the reverse effect, increasing "working-class contempt for both the middle class and higher education."[106] There is, for many, a fear that "an educated kid could morph into Them, the boss-type people many working-class folk have learned to despise throughout their clock-punching lives." As a result, Lubrano found in more than one hundred interviews that "straddlers"—people from working-class backgrounds who have made the move into the middle class—were "liable to feel hopelessly alienated from those who raised [them]."[107]

In contrast with the middle-class tendency to focus on "careers," members of the working class are more likely to have "jobs" that are starkly distinguished from their family lives. Lamont found, for example, that the working-class men she interviewed held an "overriding commitment to private life."[108] In fact, a range of research indicates that working-class men and women generally put family above work and find greater satisfaction in family than some members of the middle class, in part because family is the realm of life in which they can be safe and in charge. As Gorman noted, "working-class parents think there is a higher calling for being a parent that those with a socioeconomic advantage do not appreciate."[109]

Middle- versus Working-Class Practices of Democracy

Divergent approaches to democratic social action are associated with each of these class cultures. Arising from the penchant of the middle class for extended rational dialogue and its veneration of individuality are overlapping visions of what I call *collaborative* and *personalist democracy*. In contrast, a preference for what I term *democratic solidarity* emerges out of working-class commitments to mutuality and tradition, the embodied nature of work, and limited resources of time. In important ways, these democratic practices represent transformative versions of the daily practices of each group: what I described previously as the collaborative association of the former and the organic solidarity of the latter.

In this section, I turn back to history, summarizing the ways these different practices emerged in each class. The chapters that follow flesh out this sketchy discussion. With respect to the middle class, I focus on turn-of-the-century collaborative progressives, especially Dewey, and on personalist intellectuals and educators in the 1920s and 1960s. For the working class, I look to Saul Alinsky, the dominant formulator of community-based democratic solidarity whose organizing work began in the late 1930s, and to the writings of organizers who came after him.

Progressivism as Middle-Class Utopianism

As the nineteenth century ended, the middle class suffered from a discomforting sense of uncertainty in a world that seemed increasingly morally and materially adrift. Old cultural commitments, old understandings of the economy—everything seemed unmoored. These general fears were magnified by titanic struggles between labor and capital that waxed and waned throughout the last three decades of the 1800s and, at times, seemed to threaten the very fabric of social stability in America. At moments, it could seem like "the United States faced a mass rebellion."[110] At first the wrath of the nation and of the middle class fell mostly on workers. Although violence in the labor struggles of these years was often initiated by employers, it was workers who suffered the most profound loss of credibility. Years of conflict led to "the impression that the nation's labor elements were inherently criminal in character: inclined to riot, arson, pillage, assault, and murder."[111] In response came decades of brutal antilabor campaigns by employers, the courts, and the state.

Over time, however, large sections of the middle class, along with much of the rest of the country, became almost equally uncomfortable with the enormous wealth and dominating power of the captains of industry and their expanding corporations. They were repelled by the tendency of the "upper 10" to treat their workers like machines and especially roused to anger by child labor and the apparent disorder and incredible poverty of growing slums in the cities.[112]

Together, these conflicts and concerns produced revulsion on the part of many middle-class people for both owners *and* workers. Both sides seemed like children: unable to get along, to cooperate as rational people should—as the middle class did. A central goal of progressive reforms in the early decades of the twentieth century, then, was finding a resolution to what they perceived as an unnecessary and destructive war between labor and capital.[113]

Three relatively distinct approaches to social reform emerged among middle-class intellectuals and policy makers at the turn of the century: what I call *administrative, collaborative,* and *personalist* progressivism. These visions reflect, in part, divisions between managers embedded in the hierarchical structure of social institutions, more independent professionals who often found their strength in association, and artists and independent intellectuals searching for cultural reconstruction and opportunities for self-expression during the machine age.[114] The differences between these three (loosely defined) groups did not constitute a fundamental fracture of the middle class, however. Managers, professionals, and artists, for example, were often raised together in the same families, imbibing the same middle-class practices. In fact, I will argue, ironically, that key goals of the administrative vision are actually well served by personalist and collaborative pedagogies, even though these pedagogies were overtly constructed in resistance to bureaucracy.

In the simplest sense, bureaucrats sought methods for managing recalcitrant workers, while relatively independent professionals were more inclined to envision a social democracy that embodied either the more collaborative practices of their associations and daily work or the intimate relationships and expressive individualism nurtured in middle-class families. In Robert Wiebe's terms, bureaucrats "construed [social] process in terms of economy," seeking to "regulate society's movements to produce maximum returns for a minimum outlay of time and effort; to get, in other words, the most for your money." Collaborative progressives, in contrast, tended to explain social "process through human consent and human welfare" and spoke of "economic justice, human opportunities, and rehabilitated democracy."[115] The personalists, for their part, simply weren't that interested in the details of politics or social transformation. Society would naturally improve if most individuals were able to authentically develop in egalitarian communities.

Bureaucrats

The aims of expanding bureaucracies in an emerging corporate America were best described in Frederick Winslow Taylor's influential writings on "scientific management."[116] In Taylor's vision, management and technical experts would lay out exactly how a job was to be done, so that the only task of the worker would be to do what he or she was told. In its most basic form, scientific management involved little "science"; workers were simply pushed as hard as possible to determine the minimum time in which a particular task could be completed, and then others were pressured to achieve that speed.[117] This model appealed to capitalists, who wished to eliminate worker discretion and reduce the cost of employment, and to middle-class managers and technicians because of the respect it gave to their formal knowledge.

Sophisticated administrative progressives understood, however, that bureaucracy in a complex world could not simply consist of a static system of rules. Instead, it would necessarily embody continually "fluctuating harmonies" in response to "fluid social process[es]." This, of course, required the continual intervention of experts. Thus, bureaucrats resisted strict guidelines and rules when these restricted the scope of *their* judgment. "The fewer laws the better if those few properly empowered the experts."[118] From this perspective, the key characteristic of managerial life was the *discretion* that the middle class increasingly gained over the systems that they supervised.

On the surface, this seems like a recipe for oppressive domination of the working class, and it often took that form both on the job and in society. However, to key progressive bureaucrats like Walter Lippmann, it also provided the foundation for an increasingly popular, middle-class utopian ideal.[119] Lippmann and others hoped that through benevolent planning and management, disinterested experts could make the world better for everyone. For progressive bureaucrats, then, the new science of administration was not simply a tool for social control; it could potentially enhance the freedom and satisfaction of all. In fact, Lippmann was one of a number of former collaboratives who became proponents of such a bureaucratic, expert society, especially after World War I, as they confronted the apparently unredeemable ignorance and gullibility of the mass of humanity.[120] None of these writers ever figured out, however, how one was to identify an elite who could be depended upon to be truly objective and benevolent. Furthermore, they exaggerated the extent to which technocrats could effectively control from a distance the rich contexts and embodied experiences that dominated the working lives of the working class.[121]

Collaborative Progressives

A separate group of progressives, overlapping in complex ways with the first, sought a model for a harmonious society informed by the collaborative characteristics of middle-class culture. The collaborative form of the emerging professions which professionals used to control access to knowledge and jobs provided a crucial example of this ideal, as did increasingly more "democratic" forms of child rearing in middle-class families. If the administratives' solution to the crisis of social order was to benevolently control those from the "less civilized" upper *and* lower classes, the goal of the collaborative democrats was essentially to make everyone in society middle class.

It is important to emphasize that what the collaboratives sought was not middle-class culture as it currently existed. In fact, many were unhappy with the increasing atomization of middle-class communities and with what some perceived as their own culture's "enervating" banality.[122] They also began to associate uncontrolled individualism with the rapacious greed of the "upper 10."

Although a small number dallied with socialism, most rejected its revolutionary implications. The fact was that the current social structure of society served members of their class quite well, despite its limitations. Thus "the great majority of the middle class wanted something in between" liberal individualism and socialism. In response, prominent intellectuals developed a vision of a society grounded in what I am calling collaborative democracy. And starting in the 1890s, in scattered examples across the nation, "middle-class men and women began to create real versions of their utopia in the controlled, contained environment of small communities."[123]

Dewey, the preeminent theorist of his age, developed the most sophisticated conceptualization of this democratic ideal, but in its general outlines his vision closely resembled models developed by many other progressive intellectuals, activists, and religious leaders.[124] For Dewey, authentic democratic practices encouraged individual distinctiveness amid joint action.[125] Participation in group action should nurture individual perspectives, not suppress them, as long as they served the shared aims of society. In Dewey's famous Laboratory School, described in more detail in Chapter 2, for example, middle-class students were given many opportunities "to get from and exchange with others his store of information," and "conversation was the means of developing and directing experiences and enterprises in all the classrooms."[126] In good middle-class fashion, the children learned to collaborate by engaging in dialogue with each other and consciously planning their activities, drawing from the unique capacities of each participant. Similarly, in his writings Dewey consistently emphasized the importance of allowing individuality to express itself within collaborative action with others. This, then, was the utopian vision of middle-class champions of collaboration: a society in which citizens might maintain their unique individuality and yet escape social isolation, overcoming the banality of their lives by working together to solve common problems and create a better world for all.

Like other progressive democrats, Dewey saw "the emerging and professional elements of the middle class as the preferable historical agent" of social change.[127] Although the practices of everyone in society needed to be improved, it was the middle class that was closest to the ideal. Even the "radical" writings of pre–World War II "social reconstructionists" like George Counts, which went the furthest in acknowledging the problematic positioning of middle-class intellectuals vis-à-vis the working class—promoting socialist solutions to economic inequality and accepting the necessity of conflict in wresting resources away from the privileged—contained only hints of a coherent critique of Dewey's fundamentally middle-class vision of democratic engagement.[128]

While the bureaucrats at least implicitly accepted divisions between classes, the democrats rejected social classes as products of faulty practices and misunderstandings.[129] More generally, underlying the collaboratives' vision was a firm conviction that aggressive social conflict (as opposed to restrained discursive disagreement) was unnecessary. Although many supported the right of collective action on the part of aggrieved workers, then, they generally envisioned this on the model of rational cooperation, not, as unions often did, as a zero-sum war over limited power and resources. And unlike bureaucrats, who relied on new systems of control as sources of order, the collaboratives looked often uncritically to education as the key force for transforming "others" into discursive democrats.[130]

Like most progressive intellectuals of his time, Dewey had little extended contact with working-class people throughout his long life. However, this aversion to aggressive social conflict was visible even in the work of Jane Addams, an enormously prominent upper-middle-class collaborative progressive who lived for decades in close relationships with the poor who frequented her famous settlement house, Hull House. She was very supportive of the value of workers' traditional culture and actually allowed unionists to operate out of Hull House. Yet she rejected the necessity for conflict between labor and capital. For example, in one essay, "Addams concluded with a characteristic tinge of middle-class condescension" that "'it is clearly the duty of the settlement . . . to keep [the union movement] to its best ideal.'" At the same time as she "praised the 'ring of altruism' in the union movement," she "chided its pursuit of 'negative action,'" emphasizing that "'a moral revolution cannot be accomplished by men who are held together *merely* because they are all smarting under a sense of injury and injustice.'"[131] They would not be engaging with each other as whole beings in collaborative dialogue. She appealed to capitalists to see their workers as human beings and not just the raw material of labor. In the wake of the national strike against the Pullman company, distressed by her inability to arbitrate a solution, she critiqued both Pullman and his workers for not engaging with each other as rational human beings, for not accommodating each other's needs and perspectives.[132] Despite her great familiarity with the poverty and struggles of the poor, then, like other collaborative progressives she objected "to that word *class*," emphasizing at one point that "there are no classes in this country. The people are all Americans with no dividing line drawn."[133] Of course, she understood that these lines *were* currently drawn; her point was that they were unnecessary. Similar perspectives were expressed across the spectrum of democratic progressive writings.[134]

Democratic progressives supported labor initiatives that fit with their core commitments. With the National Civic Federation, for example, they attempted to bring businesses and workers together in dialogue. They also promoted arbitration laws in many different states. In each case, a core blindness of these reformers was to the existence of inequality that made rational collaboration impossible. They projected their experiences as professionals and managers onto the very different realities of working- and upper-class life. As a result, their efforts to democratize American labor relations were largely ineffectual and often counterproductive.[135] (The famous union organizer, Mother Jones, described the National Civic Federation, for example, as "the biggest, grandest, most diabolical game ever played on labor."[136])

Progressive support for business-controlled company unions perhaps best illuminates the fundamental limitations of their vision of collaborative democracy. While many progressives saw company unions as a first step toward democratic worker participation, businesses accurately saw them as tools for undermining worker control and resistance. In nearly every case where company-controlled unions were instituted, the rules governing participation made worker influence quite limited. Union representatives were often actively isolated away from their fellows in an effort to reduce solidarity. In fact, the limited participation allowed by such schemes often served as tools for degrading pay and employment conditions,[137] a tendency that continues today.[138] While sophisticated progressives like Dewey and others rejected the antidemocratic aspects of systems like these, the inequities that they produced were nonetheless a natural result of a social vision that, on a fundamental level, believed that something approximating social dialogue uncontaminated by power *could actually occur* in the context of industrial capitalism. Workers had learned, in contrast, that *whenever* one bracketed issues of unequal power, those with less power suffered. An equal place at the table of dialogue, their leaders understood, was only possible when workers collectively constituted a real threat.

Personalist Progressives

In the 1910s and 1920s and in the 1960s and 1970s, a second strand of progressive, middle-class thinking showed itself.[139] Drawing deeply from the European romantics, their most important precursors in America were the eighteenth-century transcendentalists, especially Ralph Waldo Emerson and Henry Thoreau, as well as the related work of poet and essayist Walt Whitman.[140] Central thinkers of the personalist camp included mostly forgotten writers like Waldo Frank and Van Wyck Brooks in the 1920s and Paul Goodman in the 1960s.[141]

As I explain in more detail in Chapter 4, I use the term "personalist" in an effort to capture this group's dual focus on the importance of authentic personal relationships within egalitarian communities *and* on the importance of nurturing unique individual expression. Personalists were just as concerned about fostering better communities as they were about nurturing unique individuality. Like Dewey, they understood that individuality and community were two sides of the same coin, that only through social interaction can people develop their distinctive capacities, even though the *kinds* of communities they sought to create looked very different from Dewey's.

Personalist progressivism emerged most strongly in the twentieth century during eras when the economic productivity of society seemed almost unlimited. Collaborative progressives had been responding to the conflict, inequality, and social instability they saw around them: the "social question" of poverty and the failure of members of the lower class, especially, to adapt to the new conditions of industrial society. This "social question" was much less important to the personalists, in part because it seemed likely to pass away by itself as a result of the seemingly inexhaustible surpluses of modern society. Instead, the personalists focused on the challenges presented by an increasingly shallow consumerism and the all-encompassing, bureaucratic nature of modern society. While personalists acknowledged the plight of the less privileged in their society, they often also romanticized the extent to which marginalized groups were more free of the strictures of modern society than themselves.

The personalists believed that bureaucracy had systematically infected modern society, slowly eliminating coherent avenues for individuality and creativity. In the 1920s, they expressed a "pervasive concern with whether man was being transmogrified into a machine."[142] In the 1960s Paul Goodman echoed these worries, complaining that it was becoming increasingly hard to find "some open space, some open economy, some open mores, some activity free from regulation cartes d'identitie." Increasingly, society, he feared, seemed to have "decided all possibilities beforehand and [to] have structured them," becoming "too tightly integrated" and preempting "all the available space, materials, and methods" for self-expression.[143]

Personalists frequently criticized collaborative progressives for their failure to perceive this danger. Goodman, for example, argued that Dewey and other collaborative progressives of the early twentieth century had "failed to predict that precisely with the success of managers, technicians, and organized labor, the 'achieved' values of efficient abundant production, social harmony, and one popular culture would produce even more devastatingly the things they did not want: an abstract and inhuman physical environment, a useless economy, a caste system, a dangerous conformity, a trivial and sensational leisure."[144] In modern society, Goodman and others argued, people encounter each other wearing the social masks that have been provided for them. Personalists sought to transform this culture, to dissolve these masks.

Collaborative progressives had looked to the emerging practices of their own class as their key model for a better society, denigrating the "primitive" nature of working-class culture. Especially in the 1960s, however, the personalists saw the middle class itself as a central problem. An increasingly debauched middle-class culture was leaching capacities for "authentic" self-expression and interpersonal communication from society. Personalists tended to look, instead, to the very "primitive" societies that collaborative progressives had earlier denigrated for more authentic modes of interpersonal interaction and expression. While collaborative progressives reached *forward* toward a democracy that had not yet been achieved, personalist progressives reached nostalgically *backward* toward an idealized premodern past in which the strictures of daily life were much looser and in which individuals had more room for individuality.

Despite their emphasis on the past, however, the personalists were themselves drawing from key aspects of contemporary middle-class culture in their celebration of authentic personal relationships and on unique individuality—especially the middle class's focus on aesthetics, individualism, and intimate relationships nurtured in the nuclear family. The "past" they imagined was in many ways more a reflection of their present than any actual earlier historical time. In truth, then, collaborative and personalist progressives *both* sought to perfect aspects of contemporary middle-class life. They simply focused on different and in many ways opposed characteristics of their own culture. Thus the personalists were as "progressive" as the collaboratives, despite their tendency to look backward for key insights about human improvement.

The personalist progressive schools of the 1920s—which Dewey attacked for their lack of focus on collaborative practice, among other issues—and the free schools of the 1960s were almost completely populated by the children of middle-class professionals. In these schools, the personalist progressives developed often quite sophisticated pedagogical strategies for nurturing egalitarian communities of free dialogue and individual self-expression. Personalist pedagogues like Margaret Naumburg and Caroline Pratt in the 1920s and Goodman and A. S. Neill (a British educator who became popular in America during the 1960s) frequently criticized collaborative progressive educators like Dewey for their failure to fully actualize the unique individuality that collaboratives also said they valued. And they rejected the ways collaborative progressives "manipulated" children into communal practices that restricted fully free, unfettered dialogue and interaction.[145]

It is important not to overemphasize the differences between the collaborative and personalist progressives, however. In fact, the personalists were deeply indebted to the work of the collaborative progressives, especially Dewey, in many cases explicitly acknowledging this. Both groups sought to support the growth of a more truly egalitarian, democratic society. And both were deeply interested in nurturing the creativity of individuals. In many ways, then, they represented two poles of a broad continuum of democratic progressive thought.

One critical area where collaborative and personalist progressives differed quite significantly, however, was in their vision of democratic social transformation and politics. The collaborative progressives struggled mightily with the details of how a democratic society might operate and with the specific practices by which democratic governance could be made most effective. The personalists, in contrast, tended to assume that if they could solve the "individuality" problem, the challenges of a democratic society would just take care of themselves. In any case, politics and governance simply were not core interests for them.

The Triumph of Bureaucracy

It should come as no surprise that the bureaucrats largely won the battle over social structure and social reform in the twentieth century. Much ink was spilled pondering the possibilities of progressive democracy, but these speculations had only a limited effect on American society. These visions have maintained a strong influence in academia, however—especially in education—and among middle-class activists.

Democratic Solidarity: A Pragmatic Response to Oppression

In their unions and in struggles to gain community power in cities, workers developed approaches to social action and social change that diverged radically from those of the collaborative and personalist progressives. Visceral experiences of oppression and poverty as well as traditions of mutualism made it clear to workers that their only strength lay in solidarity. Not surprisingly, many found socialism and other attempts to fundamentally change the structure of the capitalist economy enormously appealing, although these ideas have mostly lost their grip on workers over the last half-century.

It is true that unions, especially, have long struggled with issues of democracy. Workers' preferences for clear leadership and group loyalty, grounded partly in a chronic lack of time and resources, have frequently short circuited broad participation. Dependent on leaders to make key decisions and to negotiate for them, the working class has often found that their leaders became detached from the interests of the collective, pursuing their own interests or the interests of a particular faction in opposition to the whole.[146] Nonetheless, distinct and sophisticated models of what I am calling democratic solidarity have been developed. Here, I look not to unions but to the approach to organizing local communities developed by Saul Alinsky and evolved by his followers. I chose this focus not only because my own experience has been with organizing groups but also because organizing groups seem to evidence a stronger tradition of democratic governance.[147]

Alinsky developed his model of organizing in the 1930s in direct response to the limits of middle-class, "liberal" approaches. For example, he attacked the preoccupation of academic sociology with "the development of consensus" and its avoidance of conflict.[148] And he explicitly rejected progressive visions of discursive democracy, complaining about "liberals who have the time to engage in leisurely democratic discussions" and "to quibble about the semantics of a limited resolution," who didn't understand that "a war is not an intellectual debate."[149]

Instead of seeking a calm, rational consensus, Alinsky pursued essentially the opposite approach. He aimed, to "rub raw the resentments of the people of the community; [to] fan the latent hostilities of many of the people to the point of overt expression."[150] He instructed organizers to "pick the target, freeze it, personalize it, and polarize it,"[151] to dramatically illuminate the underlying struggle between "us" and "them." He sought to use anger at external oppression as a tool for breaking up fractures between different groups in the community and for showing people that they had more to gain by working together.

Despite his talk of war and conflict, Alinsky was not a defender of violence, however, envisioning social action as a kind of aggressive nonviolence. And anger was never an end in itself. Instead, he sought to channel resentment about oppression into a "cold anger" that linked strategy and intelligence to emotions that could sustain action.[152]

Within his organizations, Alinsky was strongly committed to democratic governance, and those who came after him deepened this. His central tool for ensuring that organizations actually represented the interests of the people was to seek out what he called "native leaders." These leaders were not those generally chosen by middle-class progressives, the professional managers who increasingly dominated institutions in the slums.

Instead, he sought out people who were actually respected and looked to by local people. And he tried to ensure that leaders actually followed and were seen by people as following the actual interests of the community.[153]

More recently, organizing has faced the dissolution of community ties of ethnic, racial, and religious mutualism that had characterized poor urban communities up through the middle of the twentieth century when Alinsky did his most important work. In response, protégés of Alinsky, like Ed Chambers,[154] have developed new practices for recreating this web of connections. I discuss these approaches in Chapter 6.

The most important education in organizing groups takes place amid action. Leaders learn both from the modeling of skilled organizers and from the real events that they encounter in the world. The focus is on the kind of "embodied" knowledge so important to working-class culture. Established organizing groups do usually provide some formal training to their leaders as well, however, teaching a common language and core concepts of organizing.[155]

This community organizing model represents a fairly sophisticated instantiation of what I call democratic solidarity. At least in the ideal, it is a thoroughly democratic form of organization designed to foster mass action under the guidance of a relatively small number of leaders who are deeply connected to the desires of their constituencies and have the time to participate deeply in decision making. It is explicitly designed around core aspects of working-class culture in its approach to action, to power, to social ties, to tradition, and to learning. Most fundamentally, this model responds to the limited resources available to working-class and impoverished people.

Putting It All Together: Cultural Capital, Material Capital, and Social-Action Practices

Figure 1.1 loosely maps the different models discussed previously on a space defined by social capital on the vertical axis and material capital on the horizontal axis. The bounded areas represent different social classes, and the descriptive text within describes the key intellectuals and social practices relevant to each, with three different and interrelated sets of practices within the middle-class "space." While in the real world the different classes would overlap more, for the sake of clarity I have left them relatively distinct. The "Upper Class" box should also be taller. Of course, a diagram of this kind only lays out tendencies; individuals from any of these groups could be found at points in their lives across this space.

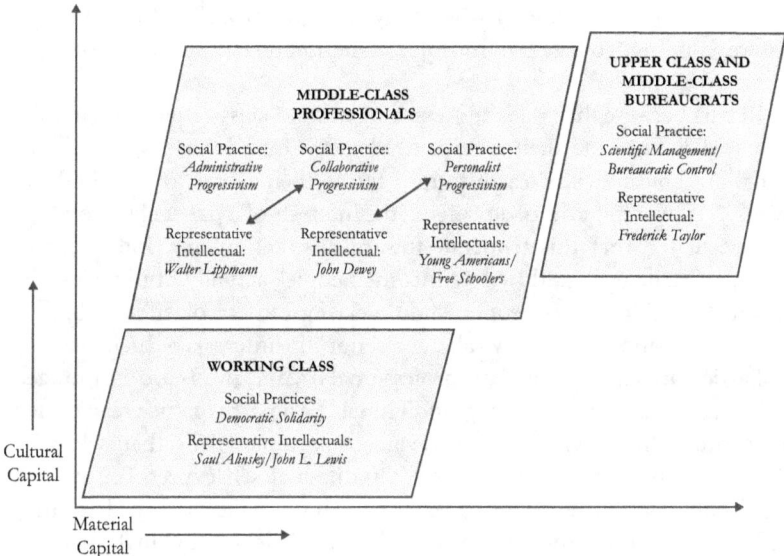

Figure 1.1 Class action practices mapped onto cultural and material capital.

Empirical Studies of Intersections between Social Class and Social-Action Practices

In Chapter 7 I provide a case study that exemplifies the class tensions discussed previously. Here, I discuss more briefly the small number of recent studies that have examined in more general terms how differences between middle- and working-class practices often play out in actual examples of collective action. The two most important analyses were conducted by Paul Lichterman and Rose.[156] In both cases, the researchers spent extensive time in groups dominated by both middle- and working-class participants.

Middle-Class Groups

Both Lichterman and Rose found that middle-class social-action organizations tend to embody the "values, ideas, expectations, and assumptions" of "successful professionals."[157] Participants are expected to conform to middle-class discourse expectations: avoiding excessive expression of emotion, depending on reasoned analysis, and making reference to "data" and expert knowledge. To participate equally, speakers need to be "comfortable with theoretical, impersonal discussion." Because they generally lack formal rules for participation, these groups generally expect people to be able to "just jump in when they want to speak," following a format resembling "college classroom[s] . . . familiar to those who are college educated."[158]

In part because the issues addressed by middle-class activists are usually only weakly linked to their specific needs, Rose found that "even the most pragmatic middle-class organizations frame their issues in broad ethical terms, . . . never in terms of advancing the interests of a particular group."[159] He speculated that this tendency toward abstraction may indicate how little the "struggles faced by low-income people" actually impinge on the "reality" of middle-class people. Middle-class groups also generally believe that they advance universally valid goals, not "the interests of their class."[160]

Participants in middle-class, professional organizations are encouraged to "continue to act very much as individuals." All participants are expected to "express their own ideas and evaluate arguments for themselves."[161] Groups often allot extensive time for individual self-expression and see it as problematic if everyone doesn't contribute. Like Dewey, then, they agree that a good community is one that can "allow individual identities and political wills to resonate loudly within collective accomplishments."[162] And like the personalists their focus on individual expression sometimes overwhelms efforts to actually engage concretely in collective action.

A range of other characteristics of these organizations also seem driven by middle-class life conditions and culture. Reflecting the often fluid nature of professional lives, for example, participation is generally understood as an individual choice, and engagement with a particular issue "may ebb and flow depending on shifts in personal priorities and interests." Because professionals are relatively free of predetermined social ties, they are continually creating "their own communities." In fact, "joining an issue organization" is one of the best ways "to meet other people who share similar concerns." Individual choice, not group history, "identifies who they are" and "establishes a community to which they belong." "Middle class politics is therefore an extension of personal development."[163]

Not surprisingly, Rose found that middle-class groups have difficulty understanding the hierarchy and suspicion about outsiders common to labor organizations. He noted that "middle-class organizations . . . find the hierarchy and formality of the union structure foreign and distasteful. Unions demand levels of privacy that are alien to peace and environmental organizations. [In contrast] these middle-class groups not only welcome but actively recruit all comers to their deliberations. Peace and environmental organizations have few if any formal rules about membership or participation. New arrivals are often asked and expected to take part in the discussion and decision-making along with people who have worked with the program for some time. Participation and equality are fundamental values."[164]

Because middle-class professionals assume that other people operate (or should operate) in the same individualistic, rational manner that they prefer themselves, they generally view "social change . . . as the product of changes in consciousness, that is, a product of education." In fact, middle-class activists often believe that people would likely act if they "'only knew about the problems being raised.'"[165] The point is not that these groups do not often seek structural changes, especially in laws, but that the mechanism for this change is often envisioned on a model of reasoned, discursive democratic education.

Working-Class Groups

The approach of most working-class groups to social action is fundamentally different. In contrast with the comparably formless character of middle-class organizations, workers' groups tend to follow established formal rules for participation and are generally organized around clearly defined hierarchies. In fact, "labor activists frequently find the meeting styles of middle-class organizations difficult and tedious." Rejecting wide-ranging dialogue about the personal opinions of individuals, they focus on pragmatic questions of action and on rituals that sustain group solidarity. As one union leader stated, the middle-class peace activists he was working with didn't "understand that it's a war out here. . . . The contrast between giving people hell at a bar over the union vote and then going to a conversion meeting where people sit around and eat cheese and sip herb tea is really frustrating. These people seem like they're from a different solar system."[166]

Those who are most respected in working-class contexts are those who most embody the core values of the working class: speaking their minds, contending, often loudly, over their commitments, and expressing the emotions behind their commitments. Eschewing abstractions, they speak from experience, often telling stories that may embody their particular perspectives but that also demonstrate loyalty and connectedness.

Membership in these groups is not simply chosen but is usually the result of a long-term embeddedness in community and family networks. Identity is something that one has, not something that needs to be found; it "comes from being accepted and known." Thus, Rose notes, "being a member of a . . . community with a good reputation defines who one is." These "close community ties" make "a clear division between members and outsiders." Trust is built over time, and newcomers are not easily allowed entry.[167]

Finally, the issues tackled by groups like unions and local community groups in impoverished areas are usually closely tied to particular community needs. Instead of focusing on universal values (although they may often refer to these), they tend to define their battles in terms of "competing interests," experiencing "their own interests . . . in opposition to the interests of others."[168] A problem is rarely seen as the result of a simple misunderstanding that can be rationally dealt with. Instead, power must be wrested from others who will generally not give it up without a fight. Win-win solutions may sometimes be possible, but experience has taught them that conflict generally involves a zero-sum game. In these and other ways, then, these organizations often embody something resembling the model of democratic solidarity outlined previously.[169]

Class Tensions

Lichterman and Rose focused on groups that especially exemplify the class characteristics I have been discussing. Even in less distinct circumstances, however, differences in approaches to social action frequently create conflicts and tensions between middle-class and working-class groups. In fact, I have watched these dynamics play themselves out in the context of community organizing efforts I have worked in over the past few years.[170] Because they have different ways of speaking, when people from different classes meet together, they often find that they can't communicate very well, misreading discursive and social cues that seem so natural to one group and so alien to the other. Furthermore, the structure of each context tends to alienate and suppress the participation of people from the other class. For example, the quick repartee of middle-class meetings can make it difficult for working-class people to get a word in edgewise, whereas the formalistic and hierarchical structure of working-class settings can seem, to middle-class members, like a tool for suppressing their individual voices.[171]

Rose summarized the differences between middle-class professional and working-class organizations:

> The middle class is prone to seeing the working class as rigid, self-interested, narrow, uninformed, parochial, and conflict oriented. The working class tends to perceive the middle class as moralistic, intellectual, more talk than action, lacking common sense, and naïve about power. Each side has a different standard for evaluating information, with the working class trusting experience and the middle class believing in research and systematic study. The result is a wide gulf in understandings of nature, sustainability, economics, and human conduct. Worse yet, working-class unions and middle-class environmentalists seek change differently. The working class seeks to build

power to confront external threats, while the middle class hopes to change people's motivations, ideas, and morality.

And he emphasized that these differences arise, in part, out of very different experiences with power:

> Different degrees of power and vulnerability are also divisive. Middle-class movements tend to have greater access to the bureaucracy because it is staffed by their professional peers. Bureaucratic processes also function through expertise and abstract rules that reflect middle-class values. The middle class tends, therefore, to have greater faith in the ability of these institutions to accomplish its goals. The working class, by contrast, is often the weakest party in conflicts and tends to pay the costs of many political and economic decisions. Its strategies reflect both this vulnerability and the interpretation of politics as a conflict about interests.[172]

Despite these gulfs, Rose argued that when they operate in isolation, class-based movements often end up "reinforcing and reproducing [problematic] aspects of society even as they work to change other aspects." For example, as we have seen, middle-class reforms have often "inadvertently served to reproduce the subordinate role of the working class in society and the economy" by placing decision-making power in the hands of experts or by downplaying the effects of inequality on democratic engagement. Working-class approaches bring their own problems, however. A tendency to focus on local interests has sometimes led working-class organizations to downplay more universalistic visions of social transformation.[173] In unions and elsewhere, a dependence on hierarchy often threatens democratic engagement. And because working-class efforts have often depended on exclusion of other, less-privileged persons from gaining access to limited resources, they can reinforce social divisions of race, ethnicity, gender, and the like.

Overall, the practices of these different groups reflect contrasting strengths and weaknesses. Lichterman found, for example, that because of their loose structures, focus on process over product, and stress on individual expression, middle-class Greens often found it difficult to act collectively or even to decide on shared goals or tactics. In contrast, the focus on solidarity in working-class groups often limits broad-based democratic participation. Both sides have much to learn from each other, if they can find a way to listen.

Social Class and Educational Scholarship

At this point I turn to a discussion of the ways these differences affect academic scholarship—focusing on the field of education. As I noted earlier, despite unique aspects, the case of education reflects progressive tendencies visible across the academic literature in the social sciences and humanities. To understand how these issues of social class have affected educational scholars and schools of education in particular, it seems helpful to look back, again, to the history of the emergence of these positions and institutions.

How Schools of Education Became Middle Class

For leading American institutions of higher education, the nineteenth century was a time of transition from finishing schools for the gentry to training grounds for children of the upper middle class. They began to shift from a focus on reproducing the classical culture of the upper class toward efforts focused on increasing knowledge, furthering social and material progress, and teaching more practical professional skills. Especially in the research institutions that became dominant forces, laboratories for natural and social-scientific investigation were founded at the same time as the dialogic practice of the seminar began to replace the didactic recitation, especially in more advanced courses.[174] Increasingly, universities took on a role as the guardians, developers, and teachers of the expert knowledge that the growing professions depended on as a warrant for their monopolies in particular areas like medicine and law. Not surprisingly, these new social-science disciplines were "imbedded in the classical," now middle-class, "ideology of liberal individualism" as well as in a strong sense of American "exceptionalism."[175]

Despite their deep embeddedness in middle-class culture, like other privileged professionals, academics tended to see themselves as floating somehow *above* any class-based interests or preferences, representing their perspectives as "objective," or "scientific." Some of the few scholars who did not subscribe to this vision made important contributions to social policy,[176] but nearly all were marginalized in the larger academic culture. And while more recent "postmodern" writings have generally rejected the exceptionalism and value "objectivism" that pervaded earlier social science movements in America, the actual discourse used in their writing has generally, if anything, been *more* "middle class" than that of their predecessors. As has been widely noted, postmodern thought has also downplayed the importance of social class.[177]

The trajectory of schools of education was somewhat more complex, deeply intertwined with the evolving structure of public schools and conflicts over the social position of teachers. After the Civil War, social class became an increasingly salient issue in the education of educators. Growing pressure to provide at least minimal schooling for all children in American society created an enormous demand for teachers. In response to a growing teacher shortage, a range of options for gaining teaching "credentials" was developed, including teaching tracks in high schools, independent teachers' institutes, and "normal" schools. Though more sophisticated than the former approaches, the normal schools that became the dominant educators of teachers around the turn of the twentieth century were more like today's community colleges than four-year institutions. The students who attended these schools "shared rather low economic status; they were, for the most part, the daughters and sons of working people," and most were women.[178] The predominance of other working-class students and the fact that normal school instructors were usually only graduates of normal schools themselves meant that these schools had limited capacity for transmitting middle-class practices.

Public schools at the turn of the century often took on many of the characteristics of factories. A broad mass of working-class women teachers taught working-class children, overseen by middle-class male supervisors. Especially in urban areas, forms of "scientific management" became extremely popular. At all levels administrators and educational scholars fought to centralize the system of schooling and to reduce, as much as possible, the discretion of "uneducated" teachers. This process was also driven by a vision of social efficiency that fit with the broader bureaucratic line of progressive thinking during this time. Students, they believed, should be trained for the kinds of jobs they would take when they left school, and for working-class children this meant learning to conform to the conditions of these jobs. Fears about working-class immigrants, especially, led progressives to "create institutions which could bring order into the lives of deviant persons and, perchance, heal the society itself by the force of example."[179]

Within more prestigious universities, however, this social efficiency approach to schooling was contested by a loosely linked group of professors promoting more "democratic," interactive, and individually responsive forms of teaching—among whom Dewey was the most important. As David Tyack and Herbert Kliebard have shown, neither collaborative nor personalist democrats had much actual impact on the structure of public schools and classrooms. Inside schools of education, however, democratic forms of progressivism became increasingly dominant.[180]

Labaree has argued that the increasing dominance of child-centered, democratic progressivism in schools of education resulted from the desire of education professors to increase their status and to see themselves as more than simply functionaries, cogs "in the new social-efficiency machine." Progressivism, he argued, provides education professors with a sense that they might contribute to the democratic transformation of the larger society, ultimately "making the reform of education a means for the reform of society as a whole around principles of social justice and democratic equality."[181]

Although Labaree was on the right path, I think he missed the most important contributor to this shift in the focus of educational scholarship: the pervasively middle-class and increasingly professional character of academic life in schools of education. And while professors were embracing democratic progressivism, their students were also becoming ever more middle class. Over the middle decades of the twentieth century, the middle class even moved to claim teaching for itself as a kind of "profession." A college degree became a standard requirement for teachers as normal schools, unable to provide middle-class credentials, either disappeared or transformed themselves into colleges and universities.[182]

Within the continuing bureaucratic structure of schooling, teachers have faced and probably will continue to face tensions and contradictions in their efforts to see themselves and act as professionals. Schools of education, however, do not have to deal with the same level of bureaucratic challenges. As their students became increasingly middle class, then, education professors increasingly structured their pedagogy around the practices most familiar to them: the practices of middle-class professionals.

The Dominance of Progressive Democracy in Educational Thought

All these developments led to the dominance in schools of education of progressivism.[183] Although most contemporary progressive rhetoric focuses on the education of individuals, the (often implicit) goal of a more democratic and equitable society is rarely far beneath the surface. And it should be no surprise that when educational scholars do speak more specifically about education for democratic citizenship, with few exceptions they look to the general model of collaborative democracy that is so indebted to Dewey, or to personalist models that focus on nurturing intimate egalitarian communities.[184] What I have described as a working-class democratic solidarity model is almost entirely missing from the field's dialogues about democratic education and empowerment.[185]

Even those few in education today who write out of at least a somewhat Marxian perspective generally look, in the end, to Deweyan democracy.[186] As Michael Apple and James Beane rightly noted, "most of the impulse toward democratic schooling" in educational scholarship today "rests on Dewey's prolific work." And although, like many other scholars, Apple and Beane acknowledged that "exercising democracy involves tensions and contradictions," they were convinced that the problem with Dewey's democratic vision is not its "idealized values" but instead our failure to fully live up to these ideals. Like nearly all contemporary progressive scholars of education, they admitted "to having what Dewey and others have called the 'democratic faith,' the fundamental belief that [Deweyan collaborative] democracy can work, and that it is necessary if we are to maintain freedom and human dignity in social affairs."[187]

There are exceptions to this pattern, of course. A few educational scholars have begun to acknowledge the limitations of discursive democracy. Critical race theory, for example, provides a very promising source of critique because of its focus on the importance of narratives and personal experience, as opposed to abstract reason, as a key source of argument and discursive engagement.[188] And a growing collection of writers outside education (especially in political theory) have been chipping away at and reconstructing the core assumptions of discursive democracy theorists.[189] Like critical race theory, this work often examines how discursive practices and strategies of social engagement differ across cultures and the ways in which a focus on "privileged" forms of discourse tends to silence those from cultures with less power. So far, however, this work remains marginal to the dominant dialogues in the educational literature, especially around student empowerment and democratic citizenship.

It seems difficult to deny that the pervasiveness of a rhetoric of discursive democracy in educational scholarship and in the classrooms of schools of education today is largely produced by the dominance of middle-class professionals. And because this cultural bias is largely unacknowledged, professional educators and educational scholars have generally seemed unable even to perceive the existence of alternative forms of democratic engagement. Thus, we have generally been unable to really critique Dewey's democratic vision or the personalist celebration of individual expression even when we acknowledge their limitations. Even in those rare moments when educators and educational scholars actually do actively promote democratic forms of education, then, we almost invariably end up embracing practices that have limited relevance, *by themselves*, to the lives of working-class students and their families.

Part II
Collaborative Progressivism

2

John Dewey's Conundrum

Can Democratic Schools Empower?

No individual ever fully "represents" a wider social and intellectual movement. The ideas that infuse a particular intellectual milieu are always appropriated and transformed, to one extent or another, in the mind of any single writer. But while I make reference to other collaborative progressives, I have nonetheless chosen in this chapter and the next to focus on the work of a single thinker: John Dewey. While every progressive intellectual would not have agreed with every one of Dewey's conclusions, overall Dewey's voluminous works contain the most complete and sophisticated formulation of the collaborative progressive perspective. If Dewey's particular vision has limitations, therefore, one is likely to find these same issues in the writings of his compatriots. Because Dewey was deeply involved in intellectual debates over democratic practice throughout his long life, his works also contain responses to the range of criticisms that emerged across the first half of the twentieth century. The fact is that almost no one over this long period could avoid framing their perspectives around these issues without engaging, explicitly or implicitly, with Dewey's vision. With respect to education, in particular, this choice seems obvious, given the enormous influence his ideas continue to have in the field.

A Representative Middle-Class Life

Even though he rarely wrote about the details of his own life, Dewey emphasized how important it is to understand the social and cultural milieu of a thinker if one is to fully understand his or her writings. He noted, for example, that "the biographical element must be stressed if we are to study motives and interests of philosophers. We must know his temperament, his personal problems, times, biases, etc... We must also take cognizance of the particular social environment of the philosopher as well as his [experience and] native sensitivities."[1] An understanding of Dewey's life is especially important for this book, since my argument is that his vision, like that of other progressives, emerged in fundamental ways out of his immersion in the emerging social realm of the middle class.

And, in fact, Dewey's upbringing and day-to-day life did resemble that of other turn-of-the-century progressives. He grew up after the Civil War in Burlington, Vermont, at a time when the painful changes of industrialism were transforming the urban landscape. As Jay Martin has shown, Burlington at the time looked much like other industrializing cities in America. In fact, Dewey's mother was deeply involved in social service in the poorest parts of the city, and "her work among the indigent exposed [her children] to the abject poverty of the lakeshore industrial area."[2] Reports about the lakeside area from that time describe "tenements where men, women, and children are herded together in violation of all the laws of decency and morality" into "haunts of dissipation and poverty,' [and] 'abodes of wretchedness and filth.'"[3] Robert Westbrook speculated that these early encounters with poverty, experienced through his mother's lens of social service, may have "shaped" Dewey's "lifelong antipathy to 'do-gooders.'" Their "altruism," Dewey felt, "betrayed a particularly subdued form of egotism." These experiences may have helped push him away from the paternalism of the bureaucratic progressives and into the collaborative progressive camp.[4]

These essentially voyeuristic experiences of working-class life were representative of Dewey's experience of the "lower" classes throughout his life. While Dewey's family was not wealthy, they were reasonably well-off, and through his mother he was connected to many in the upper class. He attended the University of Vermont, which was literally a city on a hill surrounded by the homes of the educated and the wealthy, clearly separated off from the world of the poor. He taught high school briefly (and unhappily) in Oil City at a time when few if any children of workers would have been found in secondary education. And he had little chance to experience the lives of the industrial working class during his time earning his PhD at Johns Hopkins or at the University of Michigan in his first professorial job. When he moved to Chicago, he became good friends with Jane Addams and spent extensive time at her settlement house in one of the poorest areas of the city. But there is no indication that he developed any significant relationships with local residents. And although he later argued that the general model of democratic education that he developed in his Laboratory School at the University of Chicago was relevant to children from all classes and cultures, the children he actually worked with in his school were the sons and daughters of professors and other professionals.

This pattern was largely repeated during his later life in New York City. Dewey traveled extensively and participated in many different organizations, but available evidence indicates that he never immersed himself in the working-class ways of life that his theories were, in part, meant to affect and improve.

Dewey started college when fascination with science in America was rapidly increasing. When he began as a freshman at the University of Vermont, the curriculum was slowly shifting, like other institutions of the time, from rote recitation and an established canon of classical texts to a focus on the creation and study of new knowledge. For his PhD, he attended Johns Hopkins, which was created specifically to embody new models of learning and scientific knowledge drawn, in part, from examples in Germany. At Hopkins, Dewey participated in the development of the new intellectual frameworks and scientific practices that would guide middle-class intellectuals and the emerging professions through generations to come. In fact, in his turn to an academic career, he was part of the "the first generation of university teachers of philosophy who were not clergymen."[5]

In general, then, Dewey's life was broadly representative of the experience of progressive scholars who came to maturity around the turn of the twentieth century. Nearly all significant progressive intellectuals of the time viewed the working class from a great social distance. At best, they examined or studied them. They did not attempt to get to "know" them. The only important exceptions were those, like Jane Addams, who worked for extended periods of time in settlement houses, or those, like John R. Commons, who spent time working in manual labor occupations and settings. But the settlement houses, despite efforts by Addams and others to understand and be responsive to the value of working-class traditions, were still islands of middle-class culture.[6] And progressive intellectuals like Commons were few and far between.

Creating Deliberative Scientists

From 1896 to 1904, Dewey created and directed one of the most important educational experiments of the twentieth century—the Laboratory School at the University of Chicago. Dewey created the school to give him a place to work with actual students in developing his vision of education, and it is perceived by most scholars as one of the pinnacles of progressive education. In fact, Lauren Tanner's 1997 book on the school used the metaphor of "Brigadoon" to express her desire to bring the school forward unchanged in time and her conviction that it would rival the most advanced schools today.[7] Although his time there was cut short by a disagreement with the president of the University of Chicago, the years he spent there appear to represent the only time in his life when he had significant experience with the day-to-day activities of an actual school. It was largely through his efforts in the Laboratory School that he refined his theoretical vision of collaborative democracy and democratic education. Extensive data exist about the pedagogy of the school through the detailed records the teachers kept, and as a result, it is one of the best places to look for concrete exemplars of Dewey's democratic vision. Below, I lay out the key philosophical assumptions that grounded Dewey's conception of a "scientific," democratic education before examining how he and the teachers developed and appropriated this vision in the experimental space of the Laboratory School.

Developing the "Scientific" Capacities of Individuals

Dewey believed that children learn best when they engage with tangible obstacles in their environment and come up with ways to overcome them. As they face novel situations and actively struggle to make sense of them, their efforts take them beyond routine, habitual ways of acting and nurture their capacities for imagination and experimentation. Not surprisingly, this was the core teaching strategy of the Laboratory School.

Dewey often called this process of actively engaged experimentation and problem solving "scientific." But what he meant by science in this context was quite broad. For Dewey, in the most general sense, scientific thinkers draw on past experiences and knowledge to make sense of current challenges. Amid the ongoing activity of daily life, everyday "scientists" facing some dilemma deliberate about the possible actions they could take, imagine the consequences that might arise from each possible action, and arrive at a hypothesis. Ultimately, they act and inevitably suffer or enjoy the results of their actions, learning things about the world that will inform yet future actions. A central pedagogical aim of the Laboratory School was to initiate students into this process, this ability to learn and grow in the face of new challenges.

As students move from one challenge to the next, their aims will invariably change and develop as they learn more about their situations. Because each discovery, each experimental "result," forces them to alter their goals to accommodate new information, this process nurtures increasingly flexible, creative responses to the multiple possibilities of experience. For example, children who first learn to plant individual seeds in small pots can just place each seed individually. If they try to plant a larger plot of ground, however, they will discover that this takes an extremely long time. With the encouragement of their teachers, this challenge provides an opportunity to experiment with more efficient ways to get the seeds in the ground.

In the ideal, this general abstract model of thinking should develop uniquely for each person. But Dewey worried that the uniqueness of individuals often became obscured in a society that valued conformity. All around him he saw a world in which "individuals" were "imprisoned in routine." In the traditional schools of his time, especially, he saw a tendency to force on children a "premature mechanization of impulsive activity" that destroyed the "plasticity" on which intelligent adaptation to one's environment depends.[8] He believed that, properly directed, "the intellectual variations of the individual in observation, imagination, judgment, and invention are simply the agencies of social progress, just as conformity to habit is the agency of social conservation."[9] Thus, he sought to develop a form of education that would nurture this kind of individuality.

It is important to emphasize that individuality, for Dewey, meant "a distinctive way of behaving *in conjunction and connection with*" other people, "not a self-enclosed way of acting independent of everything else."[10] The very idea of a "residual" individual outside of all associations was simply absurd to him, since an individual is inevitably different in every different association in which he or she participates. In fact, one goal of a Deweyan education was to encourage integration and balance throughout each person's different ways of being.

Dewey understood that some loss of flexibility and uniqueness was an inevitable part of learning. In fact, he argued that most learning involves the formation of what he called "habits": established, unconscious patterns of interaction with the environment. Because human beings have the capacity to pay conscious attention only to very limited aspects of our environment at any moment, there is no way to escape the importance of habits for human functioning. As a result, most of our daily activity is the result of habits. In fact, Dewey believed that habits are largely what "constitute the self," since our "conscious estimates of what is worth while and what is not are due to standards of which we are not conscious at all."[11]

While we have a tendency to think of a habit in negative terms, as with "science" Dewey used this term more broadly to refer to any established and learned way of acting. Again, however, he worried that most of the habits people learned in the traditional schools of his time or on the job were extremely destructive of individuality. "For the most part," he complained, we have given people "training rather than education," leading them to "become imprisoned in routine and fall to the level of mechanisms."[12] Factory workers, for example, were trained to repeat the same movements again and again. Because they had little or no control over their work, they became unresponsive to any creative possibilities they might encounter in their work environment. In this way, they were trained to be the "hands" for the middle-class "minds" of Frederick Taylor's vision.

In the Laboratory School and in his writings, Dewey encouraged the development of habits that were more responsive to the constantly shifting nature of every unique context. Habits promoted in schools, he believed, should *enhance* students' abilities to respond to their environment, giving them increased control over the course of their lives and the capacity to contribute to the improvement of society. Being able to drive or play the piano is a responsive "habit" in this sense.

Even the most flexible habits, however, generally treat "new occurrences as if they were identical with old ones."[13] If the novel element in a situation becomes too important to ignore, then reflection, reason, and abstraction—thinking—come into play. In these moments, everyday scientific thinkers struggle to bring a problematic situation back into harmony, a process that involves changing not only the environment but also one's habits—one's self.

For example, if a piano player encounters a key that sticks a little, she may be able to simply alter her style of play on the fly to deal with the problem—perhaps without consciously noticing this. If the key gets completely stuck, however, she will likely have to stop and figure out how to fix the key. Even after the key is fixed, however, it may still work slightly differently than it did before, forcing an alteration in her habits of playing, going forward, just as she forced an alteration in the structure of the once-broken piano. She and the piano, then, can both become changed, perhaps in quite subtle ways, as a result of her deliberative engagement with the problem of the stuck key and her return to a new harmony of practice after the problem is solved.

Dewey believed that providing children with a continual series of deliberative problems like these to solve would make students more sensitive to problematic aspects of their world, encouraging more and more frequent conscious engagement with the limitations of their society and environment. He sought not simply to foster more flexible habits, then, but also to nurture a more general tendency to grapple consciously, intelligently, with the world.[14] Children educated in this manner would gain a capacity for noticing and solving problems in the world that others less sensitive might overlook.

Collaborative Democracy

The uncoordinated activity of individuals does not by itself produce democracy. The deliberative skills and flexible habits of individuals only provide the foundation on which a truly collaborative society could be built. Collaborative progressives envisioned the emergence of a society in which everyone would be engaged, as Dewey put it, in "joint activity" oriented around common goals. Each person would coordinate their actions with those of their fellow participants, contributing to collaborative efforts to achieve shared aims. Truly democratic communities and worksites like these, the collaboratives imagined, would create a tremendous web of conscious interdependence in which "numerous and varied . . . points of shared" interest would bring people together in joint efforts.[15]

Dewey understood, however, that even in the most democratic settings groups sometimes need to act as a collective in which each participant has his or her own predefined and limited duties. A "pragmatist" and not a theoretical purist, he knew that society would always need to depend on many different forms of individual and collective action. Nonetheless, in the best of all worlds, moments of hierarchy and solidarity would take their cue from the decisions emerging from moments of collaborative democracy.

One of the key results of intelligent, democratic action would be the development of better shared habits for the larger society. Dewey repeatedly warned that one of the most crucial problems of contemporary society was that its shared habits were often left over from previous times, becoming obsolete and obstructing more productive responses to a changing environment. While Americans increasingly understand that their gas-guzzling ways are destructive to their economy and the world, for example, they have found it very difficult to alter their ingrained culture of driving. And we are still trying to adapt to the new realities of the Internet, email, and instant messaging. One of the central aims of joint inquiry in community, for Dewey, was finding ways to bring our shared habits and our environment into more harmonious balance.

Again, however, Dewey's vision of a democratic community was fundamentally one in which unique individual action becomes the "engine" of positive social change. While Dewey understood that a society with no shared habits could not operate, as a community approached his more ideal collaborative democratic model the bonds that held it together would necessarily become increasingly oriented around *conscious joint activities* to which each individual would make unique contributions.

Learning to Be Democratic Citizens in the Laboratory School

Dewey's plan was to make the Laboratory School (and ultimately schools in general) a miniature example of the kind of society he wished to promote. He sought to develop citizens who could take full advantage of the scientific discoveries and practices of the age, citizens who could use these new powers in collaborative efforts to produce an ever more democratic society. And a range of different approaches were used to encourage the development of practices of intelligent deliberation and joint action within the Laboratory School.

To initiate students into habits of "scientific" deliberation, for example, they were constantly engaged in experimentation in a wide range of different activities, including farming, weaving, and especially cooking. For example, in their book on the Laboratory School, teachers Catherine Camp Mayhew and Anna Camp Edwards described how children experimented with eggs, learning their different characteristics. "The effect of heat on albumen was worked out by first finding out the way in which the temperature of the water could be determined from its appearance—thus were worked out the scalding, simmering, and boiling points. The next step was to subject a little white of egg to each temperature for varying lengths of time—drawing thence such inferences as the following: 'The egg albumen had a very few threads in it at 140, at 160 it is jelly-like, and at 212 it is tough.' ... After these underlying principles were grasped, the work became more deductive, so to speak." As a result of experimental efforts like these, Mayhew and Edwards noted, students became increasingly "confident when confronted with the cooking of unknown foods. They knew how to discover just how tough cellulose of the new food was and the approximate amount of starch in it or of albumen. They were able to judge whether the food was to be used for flavor, for roughage, or as a source of energy. ... Such daily experience freed them from a helpless dependency on recipes, which teaching in cooking often gives." More generally, Mayhew and Edwards presented as exemplary the report of a former student who stated that in their later lives he and his fellow students tended not to "vacillate and flounder under unstable emotions" and instead would use the experimental approach they had learned to "go ahead and work out" problems in the face of emergencies.[16] This, of course, was exactly the kind of person Dewey sought to develop.

With respect to collaboration, nearly all activities in the school encouraged cooperative activity of one kind or another, even at very young ages. For example, Mayhew and Edwards reported that the four-year-olds "preferred to play alone." However, "with skillful management the climbing, jumping, running, and rolling were guided into group games." And students were taught to value the distinctive contributions of *every* member of their groups. Mayhew and Edwards told a story, for example, about "a group of children between the ages of seven and eight years, below the average in musical development" that "wrote a song which is saved from monotony by the final phrase given by a boy almost tone-deaf." In this way, "the children learned to accommodate themselves to others and to express themselves in the presence of others." As a result, "each child came to see that orderly self-direction in his activity was essential to group effort . . . The 'good' way of doing things developed in each situation, and the best order of proceeding with the activity was formulated by teachers and children as a result of group thought." More broadly, teachers continually created situations that required joint efforts of the collective to succeed. In this way, each student was given many opportunities "to get from and exchange with others his store of experience, his range of information."[17]

The students' effort to build a clubhouse together was perhaps the most paradigmatic example of this. As Mayhew and Edwards told it, the older children initially came up with the idea for the clubhouse and attempted to build it alone, but "as the work went on . . . [they] realized that what they had undertaken was beyond their own powers to accomplish, and little by little the whole school was drawn into cooperative effort to finish the building . . . Because of its purpose, to provide a home for their own clubs and interests, it drew together many groups and ages and performed a distinctly ethical and social service. It ironed out many evidences of an unsocial and cliquish spirit which had begun to appear in the club movement."[18] The clubhouse brought the entire school together around common projects requiring the creative efforts of each individual. The continual series of problems this effort created required the students' processes of joint inquiry to solve. Social problems of group difference appear often to have been solved in the Laboratory School in this fashion, engaging the students in common projects that required differences and conflicts to be overcome if they were to succeed.

Norms were slowly developed over the years of the Laboratory School's development, and the teachers reported the emergence of "a sense of security born from years of working in and with the group, a trust in the efficacy of cooperative action for the reconstruction of experience."[19] Everyone in the school shared these norms, and the teachers were ready to step in when there were problems, even going to the extent of temporarily removing students from the larger community when they could not cooperate effectively. In fact, it is important to emphasize the extent to which these activities were initiated and guided by teachers' subtle efforts, something that Dewey felt later "progressive" educators (especially the personalists) often missed. Protected by the teachers, children learned that in this microcommunity they could trust others to act in a collaborative manner on the common projects they engaged in.

This stress on cooperative activity did not mean that Dewey or the teachers had some unachievable utopian vision. Students were led to see that some people made better leaders at different times, that different children had different skills and aptitudes for different activities; and competition was not entirely outlawed. In fact, through their group projects the children discovered "new powers of both individuals *and* groups, new ways of cooperation *and* association." They explored a range of different approaches to social organization and learned how these different modes served different needs. The building of the clubhouse, for example, engaged the students in the creation and development over time of a range of different organizational strategies in order to ensure that they could effectively complete their project and organize its use after it was completed. Always, however, more hierarchical and more habitual forms of organization remained ultimately responsive to moments of collaborative democratic inquiry. The teachers stressed the importance of attending to the contributions of each participant and the effects that emanated from the actions of individuals and the group into the environment. As Mayhew and Edwards noted, "conversation was the means of developing and directing experiences and enterprises in all the classrooms . . . Each day's recitation was a debate, a discussion of the pros and cons of the next step in the group's activity."[20] In this way the process of democratic joint inquiry directed all other activity in the school.

Learning "History" in the Laboratory School

It was in the Laboratory School's approach to teaching history that Dewey's progressive confidence in the power of human intelligence was most fully embodied. Dewey's approach to history, like his vision of civic engagement, was drawn by analogy from his understanding of science. The study of history, he argued, "must be an indirect sociology—a study of society which lays bare its process of becoming and its modes of organization."[21] The chaff of history, the specific details that might confuse and distract children from the key processes by which people adapted creatively to their environment, was generally eliminated. This meant, for example, that political history was subordinated to economic history because "economic history is more human, more democratic, and hence more liberalizing than political history. It deals not with the rise and fall of principalities and powers, but with the growth of the effective liberties, through command of nature, of the common man for whom powers and principalities exist."[22] History taught in this way served a clear moral goal, as history became "the record of how man learned to think"[23] *in the manner Dewey felt everyone should*. In fact, Dewey freely acknowledged that "'historical' material" in the school "was subordinated to [the] maintenance of the community or cooperative group in which each child was to participate."[24]

As with other subjects, history was taught to children through the careful provision of obstacles. Students often explored very generalized, imagined situations through play, drama, and the material reconstruction of aspects of the conditions of earlier times.[25] Young children, for example, imagined that they were "primitive" peoples, naked and with no material possessions. With their imagined bare hands they had to overcome obstacles in their environment using the material resources made available to them by teachers. Later, students imagined they were "Phoenicians," forced by the spare conditions of their local surroundings to develop trading relations with other peoples. As they explored this "history," with the artful scaffolding of the teachers, the children invented new tools in response to challenges created by the conditions they faced. They created stone axes, bows and arrows, units of measure, new kinds of ships, and systems of symbols to expedite trading.

As the children grew older, the material of history became less general and "local conditions and the definite activities of particular bodies of people became prominent."[26] Specific political institutions and issues were left for these later years. Although children were given more detailed historical information to draw upon in their reenactments, these new stages appear to have built relatively seamlessly on the earlier ones. In the later stages, as in the earlier ones, children used the actual events of history simply as "culminating touches to a series of conditions and struggles which the child had previously realized in more specific form" through imaginative recreation.[27]

Collaborative "intelligence" of a Deweyan sort generally reigned supreme in these imaginary contexts. In one case, when an imagined tribe entered another tribe's territory during a migration, for example, "the two tribes consolidated and arranged to unite their forces, since less men would be needed to watch the sheep." The children worried that this would be difficult, and that if the tribes tried to separate, they could not figure out whose sheep were whose; they worked through this problem by means of an "examination of the character of shepherd life and the conditions and situations likely to cause difficulty."[28] The children thus took themselves as young scientific thinkers into the past, dealing with problems in a logical, cooperative way, even when they faced hostile foreign tribes. Other tribes, they often assumed, would operate under the same cultural "intelligence" they possessed, because, of course, the other tribes, if not entirely imaginary, were their classmates.

By framing the content and context of historical events in this way, however, Dewey and the teachers ran the risk that students would not understand the contingency, complexity, and unpredictability of social change. The emphasis placed on intelligent responses to natural and economic conditions as the clearly identifiable engines of historical and social change necessarily obscured the ways different cultures frequently operate under fundamentally different worldviews, constructing their environments through very different filters than that of the students.[29] As a result, the Laboratory School's focus on intelligent responses to material conditions may have made the students actually *less* responsive to the complex cultural forces that operate in the "real" world.

Nurturing Bureaucrats, Not Democratic Citizens

In fact, there was a deep irony in the Laboratory School's approach to the education of collaborative democrats that Dewey and his progressive colleagues seemed to have largely missed. In a fully democratic world of the kind imagined by collaborative progressives, students with a Laboratory School education would have been well prepared to join society as democratic equals. However, in the world as it actually was (and still is), Laboratory School practices would be most useful in arenas already populated by middle-class professionals and bureaucrats. In Bourdieu's terms, only spaces structured around a middle-class "field" would fully support the middle-class "habitus" they had developed.

The "fields" most welcoming of this habitus could be found within the white-collar departments of the emerging corporations and in the increasingly professionally directed service organizations. Many if not most of the graduates of the Laboratory School would have found their way into these white-collar spaces, since they provided the bulk of the jobs open to their generation. As a result, their "democratic" skills for collaborative action would have ended up largely supporting just the kind of undemocratic social relationships—acting as "minds" for workers' "hands"—that the collaborative progressives abhorred. Given the world the way it increasingly was, collaborative progressive education was likely to end up largely serving what had seemed like the diametrically opposed vision of the administrative progressives. Ironically, then, instead of helping to break down barriers between social classes, the collaborative skills children learned in progressive schools were likely to contribute to the maintenance and extension of the discretionary power of middle-class professionals over members of the working class.

The Aesthetic Ground of Collaborative Action

In Dewey's later writings, he worked out in more explicit terms a theoretical grounding for the democratic practices that he had developed earlier in the Laboratory School and elsewhere. In part, as I show in Chapter 4, these explanations emerged in response to the writings of personalist progressive intellectuals, amid the broader "aesthetic turn" of the 1920s. In preparation for that later discussion, it is helpful to understand Dewey's vision of the way aesthetic symbols, broadly understood, provide the common focal point for collaborative settings.

In 1927, in *Public and Its Problems*, Dewey explained that shared "symbols" provide the communal grounds for common aims and desires.[30] Although his discussion of symbols was vague in *Public*, in *Art as Experience* he later distinguished between what he called *scientific* symbols that contain generalizable, abstract knowledge which fits into a system, and more *aesthetic* symbols which allow a conscious, collaborative community to come into being. As he wrote, there, "the same word 'symbol' is used to designate expressions of abstract thought, as in mathematics, and also such things as a flag, crucifix, that embody deep social value and the meaning of historic faith and theological creed."[31]

According to Dewey, the abstract symbols of science, mathematics, and the like are designed specifically to allow the sharing of knowledge across contexts. He believed that everyone draws from abstractions like these in different ways all the time, but it is helpful to consider the most specialized form of abstraction embodied in the knowledge of practicing scientists. Scientific concepts, he argued, differ from more practical and contextual ways of understanding because they operate in an essentially imaginary world of systematic abstraction. While working carpenters, for example, use knowledge of geometry and physics, their "understanding" differs from that of professional scientists in that it is more deeply embedded within the specific activity of carpentry. Mapping out the foundation of a house using geometric principles is a fundamentally different kind of task than exploring the context-independent, abstract rules of geometry.

His point was not that scientific knowledge is better than more practical forms in any general sense. Instead, it is designed specifically to serve the particular needs of science as a field. Science requires a high level of abstraction because it "aims to free an experience from all which is purely personal and strictly immediate . . . whatever is unique in the situation."[32] In fact, the very characteristic that makes scientific knowledge so broadly available brings with it important limitations. Most importantly, the decontextualization of science renders "its results, taken by themselves, remote from ordinary experience." In other words, the very structure of scientific knowledge hides "its connections with the material of everyday life"[33] and practice. To be used, it must always be imaginatively reappropriated back into the complexities of real contexts and situations. It must be transformed from abstract knowledge into practical skills again. And this transformation requires the creative activity of human beings. A driving instructor, for example, can help you take the abstract rules of the road and the abstract principles of automobile operation and transform these into practices of driving that fit your physiology, particular vehicle, specific community, and so forth. No sane person would get in the car with someone who had only read a textbook about how to drive.

Aesthetic symbols, in direct contrast with scientific ones, are not decontextualized from experience: they *contain* an experience. Whereas scientific symbols are dry, emotionless, and dehistoricized, aesthetic symbols stir emotions and can act as access points to a common and deeply felt history. While scientific symbols are tools for the *achievement of ends*, then, aesthetic symbols *represent those ends themselves*. As in the case of science, Dewey used "artistic" to refer to a much broader world of objects and ideas than is contained in more formal visions of artistic creation. A national flag, for example, is a common symbol of shared national aspirations, even as it is interpreted differently by different citizens.

As with his discussion of formal science, his discussion of the workings of fine art gives a good sense of how aesthetic symbols work more broadly. Art, Dewey argued, does not simply "transfer" meanings from one person to another. Instead, aesthetic "communication is the process of creating *participation*, of making *common* what had been isolated and singular." In other words, artists do not communicate "messages." Instead, they create "experiences" that are understood somewhat differently by all who participate in them. In a manner similar to the "imagination" required to appropriate scientific knowledge for use in particular situations, authentic responses to works of art put energy out into the experiencing, as each individual contributes their own unique background knowledge to the act of interpretation. Through this process, "a new poem is created by every one who reads poetically."[34]

In broader, everyday terms, aesthetic symbols of different kinds provide common "meanings" that a collaborative community coalesces around. The children collaborating on the clubhouse, for example, each understood their common project in unique ways, and these different understandings affected the contributions and suggestions each made. The "clubhouse" was the common symbol that they all came together around. A broader example can be found in the idea of "freedom" during the Southern civil rights movement, which provided an opportunity for many different people to come together in a common project to transform their society. Each participant understood what they were doing in somewhat different ways, depending on their own unique history of experiences, although, as I note in Chapter 7, solidarity was often more important than individuality in movement settings. Common "symbols," then, can allow conscious communities to come into being, embodying the shared but multiplying interpreted hopes, desires, and fears of a community.

It is important to understand that common concerns, shared symbols of possibility, exist only because barriers limit a community's self-becoming. Otherwise, there would be no need to consciously grapple with them. Just as individuals only find the need to "think" if they encounter obstacles, if complete utopia were ever actually achieved the conscious community created by these shared symbols would collapse. Such harmony can never truly be achieved, however. Every movement toward a particular aim or desire necessarily creates new problems and challenges that must be addressed. Each "success" alters a community's aims, in part because of what has been learned through action. In the Laboratory School, for example, conscious communities of different sizes were continually forming and dissolving as obstacles were discovered and overcome. The school's successful effort to build a clubhouse represented not only the culmination of a range of connected efforts but also generated new ones as the school community struggled with how it could be shared and maintained. And these challenges continually required a return to joint inquiry if they were to be solved. Similarly, in the civil rights movement the achievement of some formal rights shifted the focus of many movement participants from legal to economic barriers to "freedom." In this way, the content of the symbol of freedom was fundamentally changed.

Democratic communities, as Dewey understood them, then, are continually reaching from the "actual" toward an "ideal."[35] And the ideals represented by the aesthetic symbols that bring them into existence are never (should never be) entirely achieved.

The Demands of Democratic Dialogue

> In the short run, as many historians have shown, Progressive reform of the political process narrowed rather than expanded the circle of citizenship. [And] Dewey and most Progressives . . . failed to acknowledge this process of exclusion . . .
>
> New immigrants and African Americans were consigned to the margins, their capacity for assimilation dependent on their slow progress, their citizenship claims contingent.
>
> —Shelton Stromquist, *Reinventing "The People"*

The practical demands of the kind of democratic dialogue Dewey envisioned were enormous. It took literally years, even within the focused environment of the Laboratory School, to initiate children into the kind of habits Dewey wished the entire society to embody. Even the middle-class students in the Laboratory School, then, had a great distance to go before they were fully "qualified" to be democratic citizens in the manner Dewey desired. The challenge involved in bringing the children of the "lower" classes "up" to this level would have been orders of magnitude greater, since members of the middle class were much closer to the ideal than children from working-class families.

Dewey himself was no racist or classist. He valued a multicultural society[36] and wrote approvingly about the wisdom of those who had not been contaminated by the narrow educational institutions of his time. Furthermore, he and other collaborative progressives were critical of their own class as well of others. While the democratic vision of the collaborative progressives was rooted in the cultural practices most familiar to middle-class professionals, then, progressives sought to build on and improve these.

Nonetheless, implicit within Dewey's vision and often quite explicit in the writings of other turn-of-the-twentieth-century progressives was a critique of the limitations of cultures whose practices were most distant from their ideal. From the "higher" perspective of collaborative democratic intellectuals, the primitive cultures of the mass of poor and working-class people in America made them appear much like children. As Shelton Stromquist and other recent scholars of progressivism have shown, collaborative democrats argued that those from lesser cultures needed to be taken care of until they grew into adulthood by internalizing more advanced practices of democratic dialogue. Michael McGerr noted that,

> true to their sense of compassion, the progressives turned to segregation as a way to preserve weaker groups such as African Americans and Native Americans, facing brutality and even annihilation. Unlike some other Americans, progressives did not support segregation out of anger, hatred, and a desire to unify whites; but they certainly displayed plenty of condescension and indifference as well as compassion ... Progressives fairly readily accepted the inequitable arrangement of segregation. They did so because usually there were worse alternatives ... Protected by the shield of segregation, the fundamental project of transforming people could go on in safety. But the cost was great.[37]

The ignorance and practical infancy of the lower classes was a crucial impediment to true democracy.

The incessant focus of collaborative progressives on education as a solution to what they called the "social problem" was to a great extent a result of these worries about the limitations of the social practices of the lower and, to a lesser extent, the upper classes. A key source of the endemic racism of so many progressives was the belief that African Americans, especially, were so far from their desired cultural model as to be an almost hopeless case in the short run. Dewey, himself, was adamantly opposed to the segregation of different groups.[38] But many progressives argued that the best solution was to give "them" their own segregated communities where they might slowly learn the practices that would eventually allow them to fully join a truly democratic republic. On a fundamental level, the vision of collaborative progressives was of a world filled with cultures that needed deep social surgery if they were ever to achieve a sufficient level of democratic capacity. No wonder they often despaired about the possibilities for change.

In my discussion of working-class visions of democratic solidarity in Chapter 6, I argue that the working-class model of empowerment developed by Saul Alinsky, what he called "community organizing," treats people as more ready, *as they are*, for full political participation in the democratic polity. Instead of asking people to fundamentally change "who" they are, as in the Laboratory School, Alinsky stripped down what was needed for effective democratic engagement to the bare essentials required to contest inequality. While effective leaders must learn many public skills, the community organizing model assumes that their broader social selves can remain largely unchanged. Alinsky understood that the effects of participation in social action would necessarily affect many parts of people's lives, but he accepted more of a fragmented, split selfhood than Dewey. In part, this likely reflects the realities of working-class life, described in Chapter 1, where workers leave home to enter work spaces controlled by others, often compartmentalizing "work" away from "family." As a result, community organizers are often, partly out of necessity given their limited resources, more respectful or at least tolerant of the cultural practices that different groups bring with them to a struggle.

Avoiding Conflict

Collaborative progressives believed that if we could just learn how to sit down and talk reasonably together, we would be able to work out our problems. And, in fact, both the pedagogy of the Laboratory School (e.g., in the example of tribes meeting for the first time) and Dewey's larger theoretical and political projects tended to downplay the necessity of social conflict. As C. Wright Mills argued, Dewey's "model of action and reflection serves to minimize the cleavage and power divisions within society, or put differently, it serves as a pervasive mode of posing the problem which locates all problems between man and nature [e.g., the broad social/material environment], instead of between men and men."[39] While Dewey sometimes acknowledged that aggressive social conflict could be productive, he generally argued that it was not, ultimately, *necessary*.[40] If people would only struggle together against shared obstacles, as the students had in their efforts to build a clubhouse, the illogical, obsolete habits that distort our relations with our environment and with others would be reconstructed without conflict *among* people. In a celebration of the American frontier, for example, Dewey noted that "when men make their gains by fighting in common a wilderness, they have not the motive for mutual distrust which comes when they get ahead only by fighting one another." In his own time, however, he worried that "instead of sharing in a common fight against nature we are already starting to fight against one another, class against class, haves against have nots."[41] It is just such an inclusive spirit, arising as a transparent byproduct of collaborative engagement with shared obstacles, that Dewey sought in his writings and that he and the teachers promoted in the Laboratory School.

This requires, as Mills argued, however, a belief in "a relatively homogenous community which does not harbor any chasms of structure and power not thoroughly ameliorative by discussion,"[42] a complaint echoed by many other Dewey interpreters.[43] Alan Ryan, for example, also noted that "Dewey's philosophy was almost in principle antipathetic to the adversarial system in politics . . . It was the role of brute power in political life that Dewey could never quite reconcile himself to."[44] With few exceptions, Dewey downplayed the necessity of social conflict, emphasizing, instead, the possibilities inherent in cooperation and the dangers entailed in the use of force or violence. And this vision of a world without social and material conflict was central to the hopes of collaborative progressives more broadly. As I noted in Chapter 1, even Jane Addams could not accept the perspective held by labor leaders of her time that contesting inequality in society and creating opportunities for equal participation in American democracy would require an essentially endless war between the "haves" and the "have-nots."

The Failure of Collaborative Progressivism

Children in the Laboratory School learned in a cooperative environment that was deeply separated from the social realities of the "mean city" of turn-of-the-century Chicago at the same time as they learned a history designed to explain the development of their own cooperative world and not the complex, unpredictable society outside. They operated in a sheltered space structured carefully to reward collaborative activity, a community where, over a long period of time, students built a set of common norms of action and trust. They did not have to cooperate with people who were fundamentally different from them, who occupied differential positions of power or who represented fundamentally different interests and cultures. The teachers carefully constructed obstacles in their social and natural environment that lent themselves to scientific inquiry and Deweyan intelligence. Students' relative isolation and innocence about the often illogical and oppressive nature of the world beyond the school was surely exacerbated by the fact that most students came from privileged professional and academic, almost certainly white, families.[45] It should not be surprising, then, that Mayhew and Edwards, who maintained contact with a number of the Laboratory School children, reported that when students left the school "society" brought "both shock and conflict to a young person thus trained . . . His attempts to use intelligent action for social purposes are thwarted and balked by the competitive antisocial spirit and dominant selfishness in society as it is."[46]

Dewey and the teachers created a middle-class utopia that reflected the desires and hopes of the collaborative progressives. They proved that it was possible for a small community to operate in the manner he hoped for the larger society. But Dewey failed to show that the practices students learned would actually equip them to foster social change in the world beyond the school. On the contrary, the little evidence available shows what one would expect. Practices designed to work with people who are certain to cooperate with you are not much use when you meet those who are not so friendly. More generally, Dewey's vision in his writings and practices in the school reflected the broader tendency of collaborative progressives to ignore the deeply imbedded inequality, oppression, distrust, and limited resources of the actual society of the time.

The Laboratory School was probably the best concrete instantiation of the democratic dreams of the collaborative progressives, and it represented both what was most inspiring and what was most impractical and troubling about this vision. It embodied both the sophistication of the collaborative progressives' overall vision and the unlikelihood that the wider world would ever take on significant characteristics of its isolated society.

3

John Dewey and a "Paradox of Size"

Faith at the Limits of Experience

The previous chapter explored a range of challenges created by the collaborative progressives' commitment to reasoned, joint dialogue as the central practice of a democratic society. The pedagogy of John Dewey's Laboratory School exemplified key limitations of this approach in the real world, preparing students for the world as Dewey and other collaborative progressives wished it had been, not as it really was.

This chapter examines one additional problem with any effort to make intelligent collaboration the core organizing practice for an entire society. This is the problem of *scale* or, as Jane Mansbridge has termed it, a "paradox of size."[1] More specifically, I argue that the theoretical model underlying the democratic vision of the collaborative progressives was derived from relatively small face-to-face settings and had limited relevance to social spaces like states, cities, and neighborhoods. Extensive empirical evidence indicates that organizations oriented around the rich, collaborative, democratic practices Dewey and his fellow intellectuals most valued are ineffective on a broad social scale.

I do not mean to critique Dewey or the other progressives of his time, in particular, for failing to solve this paradox. In different forms the challenge of scale for democratic practice has bedeviled political thought for centuries in the work of thinkers ranging from Aristotle and Plato, to Montesquieu, Rousseau, and de Tocqueville, to more recent writings by Walter Lippmann, Norberto Bobbio, and Robert Dahl.[2] Instead, my complaint is that Dewey and other progressives assumed we should act (and educate) as if we *could* solve this paradox.

Some aspects of this critique of the collaborative progressive vision are not especially new, but its implications have often been ignored, especially in the education community. During the 1920s a broad group of administrative progressives developed a wide-ranging "realist" critique of collaborative progressivism that focused, in part, on the "paradox of size." The most famous and sophisticated of these critiques was written by Walter Lippmann, a former member of the collaborative progressive faction who had become disenchanted during his experience writing propaganda during World War I. In *Public Opinion* and *The Phantom Public*, Lippmann used many of Dewey's own assumptions to show that the idea of a self-ruling, democratic society on the collaborative model was unreachable.[3] It was impossible, he argued, for individuals in a complex modern society to get access to or retain the kinds of detailed information necessary for them to make intelligent, informed choices. And even if they were, somehow, able to access and understand this information, there was no clear way that a large population could effectively collaborate on concrete actions in the way the collaborative progressives imagined. Such collaboration was only possible on a small scale. As a result, Lippmann argued, modern societies must mostly relay on experts to make decisions for an ignorant and ill-organized mass of common citizens.

While Dewey disagreed with the final antidemocratic implications of Lippmann's work, he acknowledged the importance of Lippmann's challenge, noting in a review of *Public Opinion*, for example, that "Lippmann has thrown into clearer relief than any other writer the fundamental difficulty of democracy."[4] In fact, Dewey wrote *The Public and its Problems* as a direct response to Lippmann and his fellow realists, seeking to show that their findings did not rule out the possibility of collaborative, intelligent democracy on the scale of a nation or even the world.

In *Public*, Dewey's only substantial work of political philosophy, Dewey tried to concretely explain what such a society would look like, imagining a world oriented by intelligent democratic dialogue about difficult questions, supplemented by less democratic administrative "states" to deal with noncontroversial issues. He envisioned the emergence of a "Great Community" in which myriad local spaces for rich face-to-face democracy would come together somehow, forming an enormous, society-wide space for collaborative, scientifically informed decision making.

In this battle between Dewey and Lippmann at the end of the 1920s, arguably the two most sophisticated and thoughtful proponents of their respective factions, hung the intellectual hopes of the collaborative progressives. I argue here that Dewey decisively lost that battle and knew he lost it, even though he and most collaborative progressives who followed him in the years after never came to grips with that failure.

I lay out Dewey's understanding of the challenges a collaborative approach to democracy needed to solve in the modern world, supplementing this with more recent evidence about the limitations of collaborative democracy as a governing strategy on a broad scale. I show how and why Lippmann's argument decisively won the day. Then I turn to an examination of the reasons why, even in the face of overwhelming evidence that it was unworkable, Dewey and other collaborative progressives clung, nonetheless, to their vision of democracy.

A core motivation for the apparently "unpragmatic" stance of Dewey and his colleagues was the challenge of "cultural lag": the fact that it takes so long for the shared social practices of a society to catch up to environmental or other cultural changes that make these practices obsolete. For example, we still live in a culture addicted to gasoline even though the economic and political costs of this addiction have become so visibly prohibitive. Because of cultural lag, teaching people *non*collaborative practices that might be more effective for challenging inequality in the here and now would also make it less likely that society would ever be able to achieve the kind of collaborative utopia that they so much hoped for in the future. In other words, in the service of their vision they were willing to disempower the working class, among others, at the time in hopes that a collaborative society would somehow be proved possible in the years to come.

"Publics" and the Symbolic Transformation of Distant Events

In *The Public and Its Problems*, published in 1927, Dewey argued that prior to the industrial revolution's explosion of technology and scientific knowledge, connections between local communities and the world beyond had been extremely limited. Village and town residents were largely isolated. Their limited "outside" knowledge, however, was balanced by a deep understanding of their local environment that modern citizens have largely lost.

Increasingly, however, Dewey argued that individuals and groups in modern society feel the consequences of distant action without either understanding the source of these consequences or having the power to act to prevent them. The new media of newspapers (and, more recently, television and the Internet) does not help solve this problem, since it generally transmits only isolated bits of information without supplying a deep understanding of the connections between actions and events. With industrialization, then, Dewey argued that we had entered the age of what he called the "Great Society," in which nearly every individual in the world had become deeply and complexly intertwined with each other. This Society, he warned, "is not integrated . . . , [and] existing political and legal forms and arrangements are incompetent to deal with [it]."[5] Recent analyses by David Harvey, Ulrich Beck, and a myriad of others grappling with the exploding challenge of globalization indicate that our current situation remains as challenging, if not more, than the one Dewey described over a half-century ago.[6]

Our increasing technological powers have had a paradoxical effect, then. On the one hand, science has brought an incredible increase in human abilities. On the other hand, however, the long-distance relationships that these technologies generate have "invaded and partially disintegrated the small communities of former times."[7]

To begin to get a handle on how the challenge of the Great Society might be grappled with, Dewey defined what he called a "public." A "public" is comprised of any group of people who are affected by the unintended consequences of the actions of others. Because there are many ways that actions can impinge upon different collections of individuals, there is not a single public in this vision but a vast and overlapping collection of many different publics. For example, a "public" has been created in communities across the Northeastern United States because of acid rain created by drifting pollution from Midwest power plants.

States and the Limits of Administration in a Democratic Society

Dewey discussed two key social entities that might address the governance challenges of a collaborative democratic society: states and the Great Community. The entities he called states, he imagined, could provide an institutional structure for administering the decisions collaboratively arrived at by each public. The Great Community, for its part, would bring the myriad deliberations of its publics together, somehow, in a society-wide system for collaborative decision making.

Each public would require the administrative structure of a state, an essentially regulatory agency to handle issues related to the particular constituting challenge of each individual public—pollution, public health, and so on. A state would be required, for example, to enforce emissions limitations set on Midwestern power plants to stop acid rain.

States in Dewey's description are quite restricted in what they can do, however. They can be "concerned [only] with modes of behavior which are old and hence well established."[8] In other words, *states only deal with what is both predictable and generally agreed upon*. They can enforce rules arrived at through intelligent democratic collaboration, and they might be able to promulgate regulations based on those rules. But *new* challenges and innovations, by definition unpredictable and often controversial, remain outside of a state's legitimate area of control. Dewey argued, in fact, that "about the most we can ask of the state is that it put up with . . . [the] production [of new ideas] by private individuals without undue meddling." When states start trying to address the unpredictable, they start doing the work that, in Dewey's vision, should be conducted through democratic collaboration. Innovations become the providence of a state only when they cease to be innovations, when the consequences they produce are well understood and agreed upon, becoming "an article of common faith and repute."[9] In other words, states take over when democratic problem solving is largely concluded. In the same way that established habits free individuals from paying conscious attention to unproblematic aspects of their environment, then, states free publics from grappling with issues and problems that are no longer worth the investment of limited resources for intelligent inquiry and action.

States also can make both democratic and nondemocratic engagements in a public more manageable. For example, stable and established criteria for contracts reduce the complexity of face-to-face interaction, providing predictability in specific arenas of interpersonal interaction. Agreed-upon rules for participation in a group similarly can make it more likely that everyone will be able to contribute equally to a collaborative effort. Like established habits, laws and rules like these allow individuals and groups to focus their limited attention on aspects of their interaction that most require rich creativity. They free limited resources of conscious attention "to deal with new conditions and purposes."[10]

States are "formless," however, without officials who can represent the interests of the publics they are created to protect. Laws, for example, are unenforceable without judges to establish what they mean. Like the abstract scientific knowledge discussed in Chapter 2, an abstract law requires an actual individual, in an actual situation, to give it concrete meaning in every specific case. Thus, the selection and guidance of officials is a "primary problem of the public."[11] And because all publics are made up of collections of unique individuals, officials can, at best, only interpret what their public as an *imaginary* collective might think.[12]

Despite different theoretical leanings, the recent work of Ernesto Laclau is helpful here. Like Dewey, Laclau argued that collectives like publics require representatives to "speak" for them because they "are" only tenuous and never entirely formed conglomerations of individuals and groups. Because there is no concrete "voice" of such a public that can be definitively located, when a representative speaks, Laclau argued, he or she cannot avoid "transforming the identity of the represented . . . There is a gap in the identity of the represented that requires the process of representation to fill it."[13]

Although Dewey's disliked "aggregated" desires and "group minds," then, there is a sense here in which the officials of a state cannot avoid operating as if something like a unitary collective exists. (And, in fact, Dewey himself often grammatically treated publics as if they were collective subjects with opinions and desires of their own.) In essence, public officials are accountable to a collective vision that *does not*, strictly, *exist*. This unavoidably creative and subtly dominating aggregation of myriad views into singular laws and policies is only palatable in Dewey's model because the issues involved or the differences likely to emerge between individual interpretations are seen by participants as relatively unimportant. Allowing officials to treat these issues *as if* there were unanimity allows the members of the public to direct their multiple perspectives to issues and projects more deserving of conscious dialogue and activity.

By developing the concept of the state, Dewey was trying to respond to Lippmann's assertion that the vast complexities of modern life were beyond the capacity of individuals to deal with. With the state he provided a space where the activity of somewhat-independent experts and administrators making decisions *for* the people seemed like a reasonable compromise. In the ideal, at least, in Dewey's states the activity of experts was not paternalistic or controlling as long as they limited themselves to the charges given them by the larger democratic community. They would simply take care of noncontroversial issues that nobody else could be bothered with enough to disagree about.

At any time, however, events or shifts in the environment can "unsettle" apparently settled issues. Traffic laws present a useful example. Although there are many differences in how individuals view any particular traffic law, these are generally not important enough for us to direct our attention to. Specific issues and events, however, can tease these differences into the open, engaging a public in conscious dialogue. Disagreements over the 55-mile-an-hour speed limit or motorcycle helmet laws, for example, reveal how much our individual understandings of driving diverge. They illuminate the complex and multifaceted ways different aspects of these issues interact with our very identities as citizens and with our ideas about, and unique experiences of the nature of freedom, responsibility, and so on. Dewey's states, then, are the repositories of tenuous compromises that can be opened up again at any time.

Because of states' extremely limited areas of legitimate power, Dewey understood quite well that, however elegant his description might be, this concept really did not provide an answer to Lippmann's core challenges. As Lippmann had shown, the number of issues on which easy and uncontroversial agreement could not be reached, and thus for which state action would be illegitimate, was vast, while the ability of individuals to understand and attend to these issues remained extremely limited.

Probably because of their broad inherent limitations, he never returned to any sustained discussion of states after *Public*. And, in fact, his discussion of states in the first half of *Public* was largely meant to provide a solid foundation for his discussion of the problems involved in developing what he was really interested in: democratic governance to deal with "unsettled" issues in the Great Community.

Competent Individuals for an Omnicompetent World

Dewey argued that part of the problem of the limited knowledge of ordinary people could be solved by having the media and other educational institutions take on more sophisticated functions. For example, the media might be able to successfully communicate information to laypersons about even quite complex topics by "freeing the artist in literary presentation."[14] Such presentations, he hoped, might allow nonexperts to understand enough to adequately participate in collaborative democratic decision making about a wide range of complex issues.

In fact, Dewey argued that it was only through dissemination and public judgment that abstract scientific knowledge could take on concrete meaning, since only "the man who wears the shoe knows best that it pinches and where it pinches, even if the expert shoemaker is the best judge of how the trouble is to be remedied."[15] Only by seeing how expert knowledge relates (or fails to relate) to everyday practical activity can one make sense of its usefulness. Actual drivers, for example, often understand better whether a particular innovation in a car solves a problem than the engineers that designed it, since engineers generally understand the parameters of their solution in abstract rather than practical and embodied terms.

Further, Dewey pointed out that most of the knowledge of nonspecialists is not learned, abstract knowledge. Instead, most everyday knowledge is "embod[ied]" in the interrelationship between people's habits and the environment, "in practical affairs, in mechanical devices and in techniques which touch life as it is lived." He imagined that a society with more sophisticated habits would necessarily generate "a more intelligent state of social affairs, one more informed by knowledge, more directed by intelligence" that would not require improvement of our "original endowments by one whit, but . . . [which] would raise the level upon which the intelligence of all operates." While contemporary citizens are not necessarily "smarter" than stone-age hunters, for example, the ability of some of us to drive cars and use computers has vastly extended aspects of the "intelligence" of our activity. In our factories, a single worker at his or her desk can control and direct a whole army of robotic assemblers. A single pilot can control a whole fleet of drones, keeping track of broad swaths of territory. Neither the worker nor the pilot needs any deep understanding about how all the systems that support their increasingly vast capabilities work. With advanced enough capacities and tools of this kind, Dewey thought it would not be "necessary that the many should have the knowledge and skill to carry on the needed investigations"; instead, all they would need is "the ability to judge of the bearing of the knowledge supplied by others upon common concerns."[16]

The problem with these responses is that, in the end, they do not really answer questions about how collaborative democracy might be initiated on a broad scale. The information that can be conveyed from a wide range of individuals about where a shoe "pinches" is far removed from the multiplicity of perspectives involved in democratic engagements with controversial issues. Only on fairly simple problems on which these differences of opinion are not really that important—whether a shoe is too tight or not, for example—is this kind of "collective" opinion likely to be adequate.

The Great Community and the Problem of Scale

How could people collaborate effectively across vast numbers of participants? In *Public*, Dewey imagined a Great Community, a process for society-wide democratic engagement and decision making. In contrast with states, the Great Community would provide a social space for creative collaboration and participation on the scale of a nation or even the world. In the Great Community each individual would be able to contribute his or her unique capacities to an intertwined collection of society-wide common projects.

For individual contributions to count as intelligent, however, Dewey was clear that each participant would need to understand, at least to some extent, how his or her specific contributions would reverberate through the myriad interconnections of other actors and contexts in society. As he noted near the end of *Public*, intelligent collaboration "*demands* ... perception of the consequences of a joint activity and of the *distinctive* share of each element in producing it. Such perception creates a common interest; that is concern on the part of each in the joint action and in the contribution of each of its members to it."[17] His point here was not that each person must be able to know in absolute detail how his or her actions reverberate out into the environment and through the responsive actions of others. But to the extent that individual actors *cannot* trace such effects, they will increasingly become unable either to perceive the nature of their own "contribution" to their shared project or to learn from the results of their particular actions. As the scale increases, each individual action increasingly becomes like the flapping of a butterfly's wings that, chaos theory tells us, somehow, in some unfathomable way, affects the weather in another country. If people don't understand the effects of their actions, they cannot intelligently adjust their future actions in response to what they discover in the now. They cannot act "scientifically." In this way, Lippmann's problem of the "omnicompetent" individual emerges again as crucial for the Great Community.

Dewey understood that the ability of individuals to trace the effects of their actions was quite limited. He noted early in *Public*, for example, that "no one can take into account all the consequences of the acts he performs ... Any one who looked too far abroad with regard to the outcome of what he is proposing to do would ... soon be lost in a hopelessly complicated muddle of considerations. The man of most generous outlook has to draw the line somewhere, and he is forced to draw it in whatever concerns those closely associated with himself."[18]

Yet this problem of omnicompetence remained at the center of Dewey's Great Community, as it did in the writings of other collaborative progressives. Why? Because the collaborative vision of a democratic society was essentially derived from experiences of interaction in small, face-to-face, local communities, emerging from these middle-class professionals' experiences in relatively small scientific, professional, and educational groupings. Dewey did acknowledge that the "local town-meeting practices and ideas" that the United States had inherited were entirely inadequate to the demands of a modern nation-state. Nonetheless, *Public* remained deeply infused with a sense of the importance of just such local structures for the development of the Great Community.

Early on in *Public*, for example, Dewey discussed the social connections that can only develop in small, tight-knit communities. In a lengthy quotation he included, Hudson spoke of a small village in which "the tidings of [an] ... accident would fly from mouth to mouth to the other extremity of the village, a mile distant; not only would each villager quickly know of it, but have at the same time a vivid mental image of his fellow villager at the moment of his misadventure, the sharp glittering axe falling on to his foot, the red blood flowing from the wound; and he would at the same time feel the wound in his own foot and the shock to his system." Only because the villagers shared a local form of life and knew each other so well, Dewey and Hudson emphasized, could they know viscerally from only a few words the experience of the accident.[19]

In the last pages of *Public*, Dewey turned back to the local again, acknowledging that the relatively abstract signs and symbols that make a conscious public possible can only come to fruition in the dialogue and contexts of small, intimate communities. We cannot, he pointed out, have a rich conversation about a shared symbol unless we share a common history and understand the unique aspects of each other's perspectives. Only in local arenas can individuals engage each other in the rich wholeness of their particular "knowledges," drawn from deeply rooted histories of interaction. Only in tight-knit communities can individuals begin to hear others and be heard themselves *as individuals*. Thus, Dewey noted, "in its deepest and richest sense a community must always remain a matter of face-to-face intercourse."[20] (He was likely influenced in these sections by the personalists he would publicly denigrate in the next few years, as I discuss in the next chapter.)

Beyond such very small-scale interactions, one increasingly enters the realm of generalization and aggregation. Agents are, at best, able to follow only the collective effects of the actions of groups on the actions and dispositions of other groups. As Hannah Arendt, who struggled with a similar dilemma, noted, "the application of the law of large numbers and long periods to politics and history signifies nothing less than the willful obliteration of their very subject matter."[21] The use of probabilities and statistical techniques such large numbers require cannot, by definition, discern the unique effects of individual actions.

Although Dewey often wrote as if the solution to this challenge rested on some development in communications technology, in fact the problem was not one of speed or accuracy of communication but of *vastness*. More communication would only make the problem worse. The Internet, for example, does not make this challenge any simpler. To solve this "paradox of size," one would need to imagine how individuals might maintain, at least to some extent, a personal relationship with each person and context affected by their actions.

In *Public*, Dewey implied that there might be some way to link the concrete experiences developed in local settings together to allow the emergence of the Great Community. But exactly how this could happen is unclear. In fact, it seemed essentially ruled out by the rest of his argument. While Dewey argued that "the Great Community, in the sense of free and full intercommunication, is conceivable," then, he only provided vague hints of what it might look like. The Great Community "can never possess all the qualities which mark a local community," he acknowledged; instead, "it will do its final work in ordering the relations and enriching the experience of local associations."[22] But what would this mean in reality? What form could it possibly take given the challenges he raised?

One possibility Dewey dabbled with during his lifetime would be to construct a society out of a federation of many overlapping local communities.

For example, while Dewey opposed any kind of socialism with a centralized, top-down government, he long supported, in general terms, some kind of "guild socialism" that would include "a federation of self-governing industries with the government acting as adjuster and arbiter rather than as direct owner and manager."[23] Similarly, in *Freedom and Culture* he spoke approvingly of Jefferson's vision of a "general political organization on the basis of small units, small enough so that all its members could have direct communication with one another and take care of all community affairs [which] was never acted upon."[24] Certainly this approach fit with his general vision of a good society as one involving a myriad of different overlapping and freely communicating associations.

The problem is that federation models like these appear to require the kind of "representation" of collective views Dewey reserved for states and ruled out for the Great Community.[25] In fact, in a recent article evidencing a number of very Deweyan commitments, Mansbridge explored the limits of this federation idea in detail, drawing on a range of empirical analyses of different organizations. Mansbridge agreed with Dewey that face-to-face communities "are the most likely to stimulate the expansion of individual capacities for solidarity and meaningful activity . . . [in which each] can come to know the others with a kind of wholeness that derives from experience with many different aspects of their lives." However, more than a half century after Dewey's *Public*, she noted that we still have not found ways to form these small organizations into larger federations with effective power. A federation solution, she pointed out, requires that complex and multifaceted dialogues conducted in local arenas be conveyed by representatives into other arenas. Even when she explored this idea in a relatively small organization—a workplace of 41 members—however, she discovered that "representatives found it impossible adequately to convey in council the arguments on both sides and the shades of meaning that different subgroup members had expressed."[26] This problem was only magnified when she examined it on a national scale in her study of the failure of the equal rights amendment.[27]

Mansbridge's work indicates that the problems of representation create enormous barriers for those who would move between the rich complexity of local dialogue into higher levels at which the activities and perspectives of myriad locals might come together into some kind of collective power. Thus, Mansbridge concluded unhappily that "in practice, the institutions that help us find a solidarity based on the encounter with another's whole self . . . are different from the institutions that organize influence in collectives that must address problems that cover wide geographical areas. The organizations that can do one of these jobs cannot easily do another."[28]

While, like Dewey, Mansbridge argued that a combination of the two approaches to action "has not yet been proved impossible," unlike Dewey and the collaborative progressives she was more willing to acknowledge that "we may have to settle for a division of labor in which citizens participate in both kinds of associations for different ends." She was not satisfied with this solution, but could see no other. Even though she agreed with Dewey that we must continue to experiment with ways to move beyond it, she understood that without pursuing this dichotomous solution in the present, accepting that for some purposes collaborative democracy is ineffective, groups may not be able to generate "the power necessary to protect against oppression or [to] influence collective decisions on the large scale."[29]

Dewey himself acknowledged in *Public* that "perhaps to most, probably to many, the conclusions which have been stated as to the conditions upon which depends the emergence of the Public from its eclipse will seem close to the denial of realizing the idea of a democratic public."[30] Others have agreed. As Alan Ryan noted, Dewey's commitment to "building 'the great community' in which the qualities of face-to-face interaction of the village are replicated across a continent may not be a fully intelligible project, and the vagueness of Dewey's account of the 'planning not planned society' may well reflect the implausibility of the project itself."[31] "Inadvertently and ironically," Robert Westbrook added, Dewey "made almost as good a case as Lippmann that the phantom public would not materialize."[32]

Teaching Utopia in the Laboratory School

While Dewey's argument with Lippmann took place more than two decades after the Laboratory School had closed, it is helpful to look at how Dewey and the teachers in the school engaged children in activities that embodied just this local and global tension of scale.

School faculty often sought to build metaphorical connections between the knowledge learned in the school and aspects of the outside world. In one revealing example, they tried to help six-year-olds understand how their home lives were dependent upon broad social systems of production and distribution. As usual, they helped the children recreate these connections largely through the imaginative power of dramatic play, exploring how wheat was distributed from the farm through the economic system. Mayhew and Edwards reported that

> it took some time for [the children] . . . to get a clear idea of the modern transportation of wheat from the farm to the big mill and the distribution

of the flour from the mill. Here again, their first ideas were worked out through dramatic play. Some of them were to be farmers, some trainmen, some mill hands, and some grocers in distant towns. The farmers were to take the wheat to the nearest small town where it could be put on train and sent to a large city mill many miles away. Here the millers would receive it and, after making it into flour, would put it on another train and send it to the grocers in the different towns where it would be sold to the farmers when they might want it.[33]

In this activity these very young students experienced a simulacrum of the workings of the larger society, of the processes of capitalism in turn-of-the-century America. Yet such a simulacrum can mislead as well as inform. In this exercise, the systemic aspects of the economy were inseparably intertwined with the interpersonal. Each link in this classroom economy involved face-to-face interaction between individuals who knew each other intimately. It was a vision of the economy transformed into a local community. Within this simulation, the problems of scale that Dewey would be unable to address so many years later in *Public* were simply made invisible.

I have already noted the problems entailed in simple generalizations about what took place in such a rich and complex context as the Laboratory School. Nonetheless, it is clear from the extensive published evidence that this pattern of focusing on essentially *local* practices of democracy, of transforming broad social spaces into essentially face-to-face experiences through dramatic play, was a common if not pervasive technique across all age groups. By engaging in such transformations, however, Dewey and the teachers implicitly prepared students for a collaborative, "scientific" Great Community that did not then, does not now, and will likely never exist. In this way, again, Dewey and the teachers seem to have misled their charges about the democratic practices that could actually work in the world beyond the schoolhouse doors.

Why Did the Collaborative Progressives Cling to Their Vision of Democracy?

The publication of Lippmann's devastating critiques led many progressives to question their commitments to collaborative democracy. However, Lippmann's books emerged at a time when progressivism in general was already "on the ropes," intellectually and politically. As I show in the next chapter, in the years after the devastation of World War I and the horror it revealed at the core of industrial society, many looked away from the hopes of progressivism in despair. Lippmann himself, as I noted, had himself been a collaborative progressive prior to the war. Dewey was one of a quite small number of this group left to defend their vision. And their exchange was quickly followed by the coming of the Great Depression. Progressive ideas would return with a vengeance, at least in some ways, with the presidency of Franklin Delano Roosevelt, but only after years of social tragedy. Lacking hope for real opportunities to put progressive ideas into action, it is not surprising that the outcome of the Lippmann and Dewey debate did not have much impact. Even in the 1930s, dreams of progressive forms of democracy, in particular, never got much traction.

This leaves us with questions about Dewey himself, however, and with the continuing commitment of progressives to his democratic vision in the many decades since. Why, even after he had pretty conclusively shown that a successful collaborative progressive democracy was more than unlikely, did Dewey fail to look more broadly for other possible options to at least supplement the collaborative vision? An easy answer would be that he had simply spent too much time and effort defending it. And there is likely some truth to this. But this answer does not give enough credit to Dewey as a pragmatic thinker.

I believe there is a more compelling justification for Dewey's laser-beam-like focus on the promise of collaborative democracy. As Dewey often pointed out, making any significant changes in the shared cultural practices of a society is an extremely difficult and slow process. Cultures are resistant to change even in the face of vast evidence that these practices have lost their effectiveness given the realities of a changing world. This, as I noted, is the challenge of cultural lag. To understand how this challenge intersected with Dewey's thinking, we need to step back a little and examine a more general tension in all pragmatic thought—what I call the "tension of aims."

The Tension of Aims

Dewey argued that the achievement of any specific aim, like the completion of the clubhouse discussed in Chapter 2, requires planning involving many different levels of aims. In the case of the clubhouse, for example, the short-term aims of current action (planning the correct pitch of the roof, deciding on the style of the door, etc.) represented steps toward the achievement of more distant aims, like the completion of the entire clubhouse itself, which were interim aims that allowed the achievement of other aims involved in how the clubhouse would be used, and so on. In the ideal, all aims are always *both* goals to be achieved in themselves *and*, at the same time, means to the achievement of other aims. Completion of the clubhouse was an ultimate aim in itself and was also a short-term aim meant to help achieve the longer-term aims of the children and clubs that would use it.

But as democratic communities move through the accomplishment of their short-term aims, they invariably discover things they had not known before. A loft room under the roof, for example, may turn out to be too noisy for groups on the main floor. And discoveries like these end up continually shifting—in small and sometimes large ways—the very nature of the aim itself. The children might, for example, decide that having more than one activity going on in the clubhouse at a time is unrealistic after running up against problems like these. These discoveries often also produce new aims ("maybe we need *two* clubhouses"). "Every genuine accomplishment," as Dewey put it, "complicates the practical situation. It effects a new distribution of energies which have henceforth to be employed in ways for which past experience gives no exact instruction. Every important satisfaction of an old want creates a new one."[34]

Aims and Tensions between Pragmatism and Utopia

As Westbrook pointed out, Dewey "was . . . constantly railing against those who were guilty of wishful thinking because of an inattentiveness to means." But Westbrook complained that it was Dewey's own "failure to constitute democracy as a compelling 'working end,' as well as the demanding conditions he set for its realization [that] made *The Public and Its Problems* a less than effective counter to democratic realism."[35]

Westbrook was not being entirely fair to the tensions Dewey was facing in his choice of "aims" in his political writings, however. The problem is that, as far as I can tell, Dewey did not emphasize these tensions himself. There actually seem to be *two* significantly different approaches to the relationship between "aims" and present action discussed in Dewey's writings. First, one can choose the kind of aim Westbrook is recommending. This aim is developed by carefully projecting the capacities and understandings one already has into a foreseeable future. Such aims bring with them some justification that they might be actually achievable. And because the final goal is linked to current efforts by a clear plan, the implications of what one discovers from efforts in the present can be mapped out into the aim itself, making the aim flexible in response to current activity. If we discover a problem with the way we have designed the foundation of our clubhouse, for example, we can predict with some certainty what the necessary changes in the foundation will imply for the final form the clubhouse as a whole can take.

The drawback of a fairly defined aim, however, is that it is limited to what one knows in some detail currently. It does not push one to strive beyond what one can coherently conceptualize in the here and now. Taken to the extreme, one could imagine such an approach significantly impoverishing human society, closing off much of the future to imaginative view, and possibly producing real despair among those who are trapped in oppressive and currently intractable circumstances.

Another approach to aims, however, was developed most fully late in Dewey's life, at the same time as he began increasingly to speak of the idea of "faith." I focus, here on Dewey's *A Common Faith*. Intelligent faith projects possibilities revealed in the present into a largely uncharted future, as Dewey did with his idea of the Great Community. As with all aims, a pragmatic faith still must shift as one learns through interaction with "the hard stuff of the world." Unlike the kind of aims Dewey had generally focused on before, however, in *A Common Faith* he noted that a religious quality can pervade "*any* activity pursued in belief of an ideal end against obstacles and in spite of threats of personal loss because of conviction of its general and enduring value." This kind of distant aim does not "depend for its moving power upon intellectual assurance or," importantly, "*belief that the things worked for will surely prevail.*"[36]

Faith, for Dewey, involved "the unification of the self through allegiance to inclusive ideal ends, which imagination presents to us . . . as worthy of controlling our desires and choices."[37] Having and engaging actively in an intelligent faith, then, can itself be a vehicle for becoming fully human. In fact, Dewey stressed that the need for faith of this kind in his "distracted age" was "urgent. It can unify interests and energies now dispersed; it can direct action and generate the heat of emotion and the light of intelligence." For Dewey, democracy itself was the most important ideal of all, embodying an aspect of the "divine." As Steven Rockefeller argued, Dewey believed that the divine "can be experienced on the level of the heart by living according to the democratic way of freedom and growth."[38]

As we have seen, Dewey had deep "faith" both in the powers of science and in the potential of human beings. In answer to his own acknowledgment that the problems he had enumerated in *Public* seemed unsolvable, for example, he noted that "one might . . . point, for what it is worth, to the enormous obstacles with which the science of physical things was confronted a few short centuries ago, as evidence that hope need not be wholly desperate nor faith wholly blind."[39] Science, he argued, had already achieved more than any had expected. Thus, it was not unreasonable to expect that science could solve these new problems as well.

Further, as Gail Kennedy pointed out, Dewey also understood that from a pragmatic perspective, the existence of a belief can itself affect the probability of one achieving one's outcomes.[40] Thus, a belief in the "ability of human nature" may be partially internal to the achievement of "freedom." If one does not have this faith in democracy, one may not act upon it, and thus there is less of a chance for achieving aims like that of the Great Community.

At the same time, however, Dewey knew that projecting ambitious aims that one has no coherent way of achieving brings deep problems, even when drawn out of possibilities clearly evident in one's experience. Without even the most attenuated plan, there is the danger that such an aim will be extremely unresponsive to what one learns in the present. Since there is no way to know either how one might reach this goal or what it might look like once achieved, it is difficult to infer what current discoveries entail for it. One may be heading in the wrong "direction" and never know it. It is possible, in fact, that a very distant aim may not be achievable at all, at the very least in any kind of conceivable time frame; in political terms, it may represent a utopia that is largely unworkable no matter how much effort one puts into it. Despite the fact that his faith was always rooted in current experience, then, pursuing the kind of aim represented by the Great Community was an immense risk, requiring a *leap* of faith.

The old accusation that Dewey was an unworldly optimist has, I think, been thoroughly discredited, not least through the work of his recent biographers Ryan and Westbrook. He knew that success in his democratic project was uncertain at best. And he held no simple vision of a dreamlike utopia, whatever one might think of his choice of aims. He acknowledged, for example, that no single democratic structure could ever achieve perfection.

Despite these caveats, however, I think the one complaint that does ultimately fit Dewey's method is that of Maxine Greene, who, among others, noted that that Dewey "lacked a tragic sense of life"—in his published writings at least.[41] There is a tension between the need to believe in the Great Community to increase the chance it would be achieved and the danger that it might not be achievable at all. Similarly, there is a conflict between providing students with democratic habits of engagement that seem congruent with the goals of the Great Community and providing students with immediately effective practices of power that might actually make the Great Community harder to achieve. Both of these tensions seem at least muted in Dewey's texts. For an experimentalist with such a distant aim and such an unwavering *faith* in science, tomorrow was always another day. It was (still is) impossible to prove that his democratic ideal was *not* possible. That is the point of faith, after all.

While Dewey was certainly right that having an intelligent faith can be extremely productive, then, he also understood that too much faith can be extremely problematic. Arendt, whose sense of the tragic infused her writings, took this point further, arguing that one of the key tenets that drove totalitarian governments at the middle of the twentieth century was the belief that *everything* was possible.[42] They saw no limits in themselves, no trade-offs that needed to be made.

If Dewey had developed more of Arendt's tragic sensibility, he might have understood more clearly the terrible risk he was taking by aiming at such a distant and improbable goal. In fact, one of the motivations for writing this book is my fear that, because he and other collaborative progressives committed so much effort to shifting the larger culture toward a form fit to operate as the Great Community, their writings may lead educators and scholars away from considering strategies more likely to empower students and citizens in the here and now. Partly as a result, we shy away from projects that are more likely to provide people with the skills to effectively make changes in society the way it is.

Rejecting the Working-Class Alternative

Dewey and at least some other collaborative progressives understood how unlikely it was that their vision of a democratic society could ever become a concrete reality. Those who most suffered from the predations of the current society, however, were not themselves but the many who lived in poverty and worked dirty, frequently dangerous, and usually only barely sustaining jobs. Thus the progressives took a chance *in the name of* those who suffered beyond their classrooms, offices, and sitting rooms. Privileged progressives were willing to give up immediate empowerment *for* the working class and poor in America in favor of the implausible possibility of an imagined collaborative democracy in some distant tomorrow.

Dewey and other democratic progressives really had no coherent alternatives to offer beyond this necessarily vague hope. They never came up with a concrete, workable structure that would allow their cherished vision of a truly collaborative democratic society to coalesce. In part, this was because they could never figure out how to solve the democratic "paradox of size." It seems unlikely that we will ever solve it. And this makes the current dominance of Dewey's vision of collaborative democracy in education and beyond among progressives and academics today enormously problematic.

Part III

Personalist Progressivism

4

The Lost Vision of 1920s Personalists

At the end of the 1920s, around the same time as he was fighting for popular democracy against administrative progressives like Lippmann, Dewey began to push back against another wing of the progressive movement that I term "personalism." In some ways, the personalist vision was even more antibureaucratic than Dewey's. As a result, to some extent Dewey was in a struggle against both more conservative *and* more radical progressives. As I show in this chapter and the next, however, this way of framing his position obscures the fact that Dewey was much closer to the personalists than he would ever admit. Dewey tended to describe the positions of the personalists as simply atheoretical caricatures of his model of collaborative democracy, but in fact the personalists developed authentic and sophisticated alternative visions of democracy and freedom that drew deeply from Dewey's own philosophy.

In terms of education, during the 1920s most schools, especially public schools, continued to be ruled by the social-control approaches of the administrative progressives. Beginning before World War I, however, a burst of activity began to coalesce around the creation of new private and more holistic "progressive" schools. While smaller than the 1960s free schools movement discussed in the next chapter—never extending beyond a relatively small number of radical pedagogues—this was still the first time that a significant group of nonadministrative progressives began to try their hand at education. Early in the decade, in the 1910s, this blooming of progressive pedagogy seemed promising to Dewey, who wrote a supportive book about some of these schools with his daughter Evelyn in 1915.[1] However, his discomfort grew with what he increasingly perceived as a diversion from his own vision of democratic progressivism. Finally, in 1929 he wrote the first of a series of stern critiques of these "progressive" schools. Part of his discomfort arose from that fact that, as Robert

Westbrook noted, "responsibility for 'progressive education' was often laid at his doorstep" whether it actually followed Dewey's recommendations or not.[2] In fact, at least some of these new progressives seemed to think they were actually following Dewey's lead. Thus, he worried that they were distorting his own pedagogical vision in the public mind.

This chapter lays out key arguments of personalist intellectuals and educators in the 1910s and 1920s. (For reasons of brevity I generally refer to this entire period as the 1920s.) I argue that many of Dewey's complaints about the personalists were unfair to the rich sophistication of their ideas and practices. I acknowledge, however, that the multiplicity of voices during the 1920s made it challenging to discern any overarching personalist theoretical framework. For that, in the next chapter, I look to the free schools movement during the 1960s and 1970s, when a more coherent framework to support a very similar personalist educational movement emerged. Overall, this chapter seeks to map out the central differences between Dewey and the personalist educators of the 1920s, at the same time seeking to establish personalism as an authentic and fully coherent alternative vision of progressive education.

Like other forms of progressivism, the personalist vision emerged out of middle-class cultural commitments. In fact, schools working in this tradition almost singularly served the children of middle-class professionals. Three key characteristics distinguished personalist from Deweyan democratic education. First, while both Dewey and the personalists emphasized individual uniqueness and egalitarian communities, the personalists were not very interested in initiating children into sophisticated models of collaboration. Second, the personalists tended to focus their energy on social development, on the emergence of healthy, happy, emotionally stable people, often assuming that cognitive development would mostly take care of itself. Dewey, in contrast, constructed sophisticated theories of individual learning meant to initiate children into a process of "scientific" problem solving. Finally, while both Dewey and the personalists held democracy up as a central value, the personalists tended to assume that a democratic society would naturally emerge if individuals were simply allowed to develop in "authentic" ways. In a general sense, then, teachers in the personalist vision looked more like therapists, while in Dewey's collaborative progressive vision they looked more like social and cognitive engineers.

Education scholars have tended to see personalist and collaborative perspectives as radically opposed to each other. In large part this is the result of Dewey's attack on personalists in *Experience and Education* (*E & E*). In *E & E*, Dewey contrasted what he declared was his authentic vision of progressive education with what he felt were the misreadings of his work by essentially thoughtless false progressives. Personalists were hardly educators at all, he complained, since they gave their students little social or cognitive guidance, following what he felt was a fantasy of "natural" development.

Personalism

Deciding on a descriptive term for this branch of middle-class progressivism in America (along with "administrative" and "collaborative") was something of a challenge. What I call "personalism" has deep roots in early nineteenth-century romantic thought both in Europe and America, but "romantic" brings with it a range of potentially misleading connotations of antirationalism and individualistic freedom. The same limitation comes with the term "liberationist," which embodies only the individualistic side of the personalist project. In the field of education, "child-centered" has a similar problem, focusing on the individual child.

Personalism as a theoretical tradition has a long history in a range of romantic, postromantic, religious, and other sources.[3] In America, personalist perspectives were visible in transcendentalism—especially in the work of Bronson Alcott and the thought of Walt Whitman.[4] In the 1930s, versions of personalism provided a key foundation for Dorothy Day's Catholic worker movement. "Personalism" as a term first came into relatively wide use in America during the 1960s, informing many of the movements of the time, including the counterculture, the free schools movement, and Martin Luther King Jr.'s efforts to overcome racist oppression in the South.[5] In its focus on communities of organic equality and "mutual support," personalism has also at times drawn from anarchist writings, including those of Emma Goldman and Peter Kropotkin.[6]

For the purposes of this volume, I use the term "personalist" fairly narrowly to describe a loose collection of social visions that celebrate the following:

- Unique individuality, often but not always understood on the model of aesthetic expression
- Egalitarian, caring communities[7]

Paul Lichterman proposed a similar definition of "personalism" in his examination of different middle-class professional approaches to "political community" in the United States: "'Personalism' refers to ways of speaking or acting which highlight a unique personal self ... Personalism upholds a personal self that lives with ambivalence towards, and often in tension with, the institutional or communal standards that surround it."[8] In his study, middle-class personalists sought out communal contexts where they could join together with others without sacrificing their unique perspectives.

Note my reference to "egalitarian" instead of "democratic" communities. While most (but not all) personalists valued democracy, they generally framed this in much vaguer terms than had Dewey and his fellow "collaborative" progressives. More broadly, personalist writings often seemed to assume that democracy on a broad scale would naturally emerge, somehow, if the right conditions for individual development in local communities were created. In a few cases—in some of the writings of 1920s personalists, for example—visions of a fully democratic society on a broad scale were actually jettisoned in favor of different forms of elite leadership at the higher levels, although this elitism was generally seen as supporting more egalitarian communities on the local level.[9]

In any case, all the educators and other intellectuals I refer to as "personalists" understood, like Dewey, that unique individuality is possible only within human communities. Theirs was not a vision of isolated or heroic individualism. The development of the right kind of egalitarian community was always central to their efforts to conceptualize and, in the case of educators, to actually nurture individual expression.

As we will see, disagreements between personalist and collaborative progressives resulted, in part, from their very different relationships to their own middle-class culture. While collaborative progressives had critiqued aspects of middle-class culture, their model of a good society generally assumed that the practices of the middle class were closest to the ones they cherished. The key challenge for the improvement of their society was the "social problem," the cultural backwardness, and material deprivation of underprivileged "others." The personalists, in contrast, tended to stress the *limitations* of the practices of their own class. Instead of containing the seeds of a more authentic future democracy, the middle class often embodied for them the worst aspects of modern capitalist society: repressed individuals trapped within the "rat race" of an increasingly bureaucratic modern society. As a result, personalist thinkers frequently looked nostalgically to what they imagined constituted more organic communities in "primitive" societies, including the remnants of premodern communities sustained by some oppressed groups in their present (e.g., Native American communities, "traditional" African American cultures). And they cherished the possibilities entailed in the potentially less restrained and controlled world of childhood as a reflection of more "natural" forms of individuality and communal engagement. As this chapter argues, however, a closer analysis reveals that the personalists also were quite dependent on assumptions drawn from their own class culture, albeit different ones than the collaboratives looked to.

Romantic Roots of Personalism

While personalist progressives in the twentieth century looked to a wide range of historical and intellectual sources, including medieval Catholic thought and eastern mysticism, a key source was European romanticism and American transcendentalism. Rousseau especially influenced later personalist educators with his vision of an education free from the contaminations of "society," feeding the romantic veneration of childhood. In Europe, romanticism grew, in part, out of the rejection of the stale banality of their society and what many saw as the emerging "machines" of bureaucracy and industrialization. Romantics championed forms of expressive individualism that embodied more organic connections with nature, and they usually conceived of unique individuality as emerging *within* community. As Michael Lowy and Robert Sayre noted, "romantic individualism stresses the unique and incomparable character of each personality—which leads logically ... to the *complementarity* of individuals in an organic whole."[10] Early nineteenth-century romantics looked to imagined organic communities of a nostalgic past for alternative visions of a good society. In Germany, Novalis, for example, opposed the "secularized, machinelike set of states" he saw in his early nineteenth-century Germany that "aimed at rationalizing all forms of economic life." Novalis argued that "in short order, the enlightened rule of efficient administration had taken over all of life, turning 'the infinite creative music of the universe into the uniform chattering of a monstrous mill.'" Like later personalists, romantic visions of an alternative social order were quite vague. They generally assumed that "in a 'true republic' ... people would be virtuous, [and] would freely and in a friendly manner cooperate with each other."[11] While the modern middle class of the twentieth century had not yet emerged, these European romantics reflected the perspective of the "middling" bourgeois of their era—especially in their search for an organic communalism free of bureaucratic control, diverging from emerging working-class oppositional cultures of solidarity.

Transcendentalism and Whitman in America

The writings of early nineteenth-century American transcendentalists reflected many of these romantic commitments. The writers most remembered today, Ralph Waldo Emerson and Henry Thoreau, tended to stress "introspection and self-reliance" over "the brotherhood of man and outer-directed behavior for the common good," but a more social vision was stressed in the mostly forgotten writings of Orestes Brownson, George

Ripley, and others—a group that Phillip Gura refers to as "associationist" transcendentalists.[12] Aspects of this vision were also reflected in the writings of Walt Whitman, especially in *Democratic Vistas*, which was a key influence on the 1920s personalists.

The new emergence of an identifiable working class in America was a key and often unrecognized factor driving transcendental thought. The labor struggles of the early nineteenth century constituted "the most fundamental reality of the material and social world" of the time, powerfully shaping "the lives and ideas of the period's elite, especially its elite radicals."[13] In different ways, transcendentalists were seeking to "redeem" this fracturing society. Brownsen and other "associationists" sought to nurture the emergence of an egalitarian society, drawing from a broad range of social thinkers and eventually creating their own model community at Brook Farm. Even staunch individualists like Emerson sought social reform, even if this was through the mechanism of self-reliance. Interestingly, those with the most explicit focus on social transformation were also those with the most direct experience with the lives of working-class Americans.[14] More "patrician" writers like Emerson and Thoreau focused on unique individuality as a solution to social conflict, while others, still privileged but with more contact with the poor, like Brownson, promoted social solutions that drew more directly from early visions of socialism.

There were also educational aspects of transcendentalism. Early on, Bronson Alcott created a "School for Human Culture" in Boston and later guided the educational components of the Brook Farm community. In both contexts, he focused on nurturing individual expression and personality through free, spontaneous engagement with other individuals and the larger environment. Like the romantics and later personalists, Alcott believed that all children were basically good. Alcott's Boston school was quite democratic, with student participation in classroom decision making, prefiguring the personalist pedagogies of the 1920s and the 1960s. His description of his Boston school sounds much like later descriptions of personalist pedagogy. In the school, one of his teachers reported that

> young people find ways enough of amusing themselves and we best leave them much to their choice in such matters; yet some slight superintendence seems becoming—some interest shown by us in their pleasures—since these exert a commanding influence in forming their tastes and characters, and cannot be safely neglected by their guardians ...
>
> "Let us play" is the privileged version of their creed, and they enter with the unction of enthusiasm into the sweet sports they love. Then they show what they are; casting all reserve aside their souls leap sunward glossy gay into the in abandonment to fancy and fun.[15]

America in the 1920s

By 1919, the ebullient [collaborative] progressive intellectuals of 1914 had become, as Walter Weyl said, "tired radicals," though, in truth, the progressives were never truly "radical."

—Walter Leuchtenberg, *The Perils of Prosperity*

For many young European intellectuals, dreams of social and material progress grounded in science and technology, already shaken by intellectual movements before World War I, were destroyed as they huddled in stinking trenches and corpse-filled shell craters. An entire generation of young men were emotionally and physically scarred or killed in a war that took on the character of senseless insanity. Instead of precursors to utopia, machines became tools of terror with the deadly spray of machine gun fire and random, anonymous death dealt by falling shells that ground the earth into a pockmarked wasteland.

The American experience of World War I was less intense, the doughboys arriving fairly late in the game. Still, the war fundamentally altered the nature of the Progressive movement in America. We have already seen how Walter Lippmann lost his faith in popular democracy after he saw how the masses could be so easily manipulated by propaganda. The discouragement of the collaborative progressives from the turn of the century was magnified by the fact that many, including Dewey, had hoped that the war would provide an opportunity for the democratic transformation of society. "Faced by the victory of political reaction and the disappointment of their hopes for a new international order," Walter Leuchtenburg noted, "they felt an overwhelming sense of their own impotence. The world seemed infinitely less tractable than it once had."[16] For the entire decade of the 1920s, most collaborative progressives let their dreams fall fallow, reemerging only after the stock market crash and the election of the Roosevelt administration during the Great Depression. Dewey, of course, was an exception to this pattern—fighting valiantly for his democratic vision throughout this entire period.

The 1920s also brought fundamental changes to American society. In the years before the Wall Street crash that led to the Great Depression, business became glorified. In fact, "to call a scientist or a preacher or a professor a good businessman was to pay him the highest of compliments." For the middle class, these were years of growing prosperity, both because of increasing incomes and because of falling prices for a dizzying range of consumer items and entertainments. At the same time, a new stratum of white-collar, middle-management workers emerged in the growing corporations and large businesses. "The nineteenth-century man, with a set of characteristics adapted to an economy of scarcity, began to give way to the twentieth-century man with the idiosyncrasies of an economy of abundance." At least on the outside, this new "man" was "aggressively optimistic" and friendlier. But he also "had less depth, was more demanding of approval, less certain of himself. He did not knock, he boosted. He had lots of pep, hustle, and zip."[17] The iconic representation of the 1920s businessman was the eponymous main character of Sinclair Lewis's *Babbitt*, whose name began to appear in "dictionaries as a term connoting a businessman caught up in almost ritualistic conformism."[18]

During these years, a broad consumer culture came to fruition in America. "Consumerism" was the other side of "Babbittry." In response to "the degradation of work and the erosion of individual autonomy in a mass, corporate culture" exemplified in Lewis's book, "people turned to leisure and consumption to find satisfaction in life." New and increasingly ubiquitous forms of advertising trumpeted the ways consumption could transform and satisfy individuals. And new production capacities, the same ones that were degrading work, resulted in a "cornucopia of material goods."[19] Although those on the bottom of the economic ladder received little of this prosperity, they also found some outlets in new options for consumption and entertainment.

With this celebration, however, came increased anxiety for many about the rapid change dissolving the institutions and cultural mores that earlier generations had depended upon. Among the middle class, a "therapeutic ethos" of psychoanalysis became widespread. In fact, "Freud became a household name" as "populizers of Freud" became "thick on the ground."[20] These populizers downplayed Freud's argument that civilized people needed to learn to control themselves and instead stressed the importance of free expression as a solution to internal struggles. For most of the population, this bowdlerized Freudianism served only to magnify their focus on consumerism and private release outside the corporate controls of a world where business reigned supreme.

As it would in the 1960s, this new middle-class prosperity led to a youthful rebellion against repressive prewar Victorian values. Before the 1920s, the problem youth for middle-class social critics were the children of the poor and uncultured. Suddenly, however, the middle class discovered that the problem youth were now their own children. At the same time, from these youths' perspectives, their elders seemed to be clinging to the inauthentic standards of a lost and self-deluded society. World War I "had proved to the youth how unreliable, if not also culpable, their elders were, forcing them to live their lives on their own terms and by their own lights . . . Devoted to truth and candor, they refused to accept the candor of their elders; they had to experience life for themselves and to snatch its pleasures at a hurried pace."[21]

In contrast with the 1960s, however, the upheaval of 1920s youth was quite restrained. While "the young were not traditionalists, . . . neither were they radical. They valued freedom of expression, but also the American capitalist system." Further, "the young were politically apathetic." As a group they were only interested in their own personal freedom; "little interested in political or economic issues, they neither pressed for change nor partook actively in political discussions." Even their sexual rebellion, Paula Fass argued, was a rebellion within and not against marriage. Aside from assertions of free expression, they ultimately accepted the fact that they were the "heirs apparent of American industrial capitalism and . . . they quite casually assumed the political and social attitudes that came with the role." In this way, "they were . . . able in their beliefs and actions to separate political conservatism from cultural liberalism, and to become businessmen and jazz hounds, Republicans and flappers." The youth of the 1920s "were optimistically and very consciously the beneficiaries of that system, and they aspired to succeed on its terms when it came time to assume their full roles and responsibilities. Neither angry nor idealistic, . . . they were at the very center of a sociopolitical order that they would soon inherit," mostly embracing their ultimate place in a world of Babbitts.[22] Interestingly, a robust and politicized student movement did emerge after the crash in 1929, but it was much more influenced by communism and socialism than personalism. As Robert Cohen noted, the student radicals of the 1930s became the "old left" the new progressives of the 1960s ultimately rebelled against.[23]

The Emergence of Personalist Progressives in the 1920s

The despair and exhaustion produced by World War I does not fully explain the emergence of personalism as a vibrant intellectual movement in the 1920s. In fact, some of the most important writings by key intellectuals of the time, including books and articles by Randolph Bourne, Van Wyk Brooks, and Waldo Frank, appeared before and during the war. It is surely no accident, however, that the two eras when significant personalist intellectual movements emerged in twentieth-century America—the 1920s and the 1960s—also represented years when the middle class experienced relative prosperity. Despite the grinding poverty of those at the economic bottom, many in the middle class experienced these as times when "necessity" and "inequality" were fading away inexorably under the pressure of enormously productive consumer capitalism. It is not entirely surprising, then, that some middle-class intellectuals turned their focus away from economic inequality and institutional reform and toward problems of individual expression and more local, organic visions of egalitarianism and freedom.

Intellectuals of the 1920s drew from an eclectic range of aesthetic and psychological theories, focusing on Freud and those influenced by him (especially Jung), new visions of expressionist aesthetics, and, of course, a range of romantic and American transcendentalist writings. They were also influenced by the writings and dialogue about anarchism circulating during these years, especially in the intellectual hothouse of New York, mingling with figures like Goldman and Alexander Berkman. In this chapter I focus on the work of Bourne, Brooks, Frank, and Lewis Mumford—usually referred to as the "Young Intellectuals" or the "Young Americans"—although others, like Paul Rosenfeld and the photographer Alfred Stieglitz were engaged in similar intellectual projects.[24]

Despite differences, personalists in the 1920s tended to coalesce around a few core values. Prefiguring fears of what the 1960s generation would call "technocracy," and following in the footsteps of earlier romantics, 1920s personalists expressed a "pervasive concern with whether man was being transmogrified into a machine." They worried "about the implications of all those bottled-up men—not just those at the very bottom" but especially "those who achieved middling success as they struggled up the corporate ladder."[25] Unlike the vast majority of their fellow middle-class citizens, they saw "Babbittry" as a threat to everything they understood as authentic about life.

Even though they railed against the banal limitations of the "Babbittized" middle-class culture, the fact is that they drew their visions of a better society from other aspects of middle-class culture, especially in their focus on the importance of individual actualization. Much like Deweyan pragmatists before them, they critiqued middle-class culture as it existed for its distortion of the utopian potential of middle-class ideals (although many would not have admitted that their own values were also middle class).

In fact, the most important group, the Young Americans, "were far more indebted to Deweyan pragmatism than they cared to admit." Like Dewey, the personalists critiqued the destructive impact of factory labor on the capacity of workers for individual development. And they understood that industrial capitalism in its current form needed to be transformed if their cultural hopes were to be achieved. In fact, Casey Blake argued that "their calls for a democratic culture of 'self expression' were intimately connected to the critique of the factory system." In general, however, they focused much less than democratic progressives on economic oppression or practical democratic reform, stressing, instead, the personal and "psychological costs of modern industrial life." [26] While they discussed the working class, the focus of the

1920s personalists was really on the plight of their own class, increasingly trapped in the banal desert of the corporate rat race.

As Bourne noted, "the modern radical opposes the present social system not because it does not give him his 'rights,' but because it warps and stunts the potentialities of society and of human nature." He believed that "the triumph of large-scale business enterprise had brought with it the massification of everyday life—the reduction of community to a herd existence." This process could only be reversed through a cultural revolution, by "rekindling the fire of imagination in a people whose work had made them docile and whose leisure was given over to commercial spectacle." Releasing the unique capacities of individuals in the context of what Bourne called the "beloved community" (a term later picked up by the civil rights movement), in small, face-to-face, egalitarian groups, was a central aim. Importantly, the self in these communities was not simply uncovered as if it were some preexisting truth but instead "crafted" at the same time as it was released from its imprisonment "behind a shell of class and social position" through aesthetic engagement and interpersonal dialogue.[27]

While they emphasized the need for critical thought along with Dewey, writers like Bourne generally "lacked Dewey's stress on rational dialogue" and careful experimental engagement. In fact, personalist visions of the actual practices or structure of better communities were usually vague and nostalgic. Like earlier romantics, the Young Americans—especially Frank—saw apparently preindustrial cultures of "blacks, Hispanics, Indians, and other outsiders as bastions" of the kind of organic communal culture they desired. Personalist writings sometimes implied that if you could just create the right kind of actualized individual grounded in the right kind of local culture, society would take care of itself.[28]

Bourne, Brooks, and Frank reconsidered their commitment to pragmatism after Dewey came out in support of World War I. Bourne's attacks on this decision are the most famous, but the others also critiqued him. While they did not reject Dewey's entire vision, they became more aware of its limitations. In general, as Blake noted, they argued "that it failed to encompass the *imagination* as well as the intellect." They thought that "the inability of pragmatism to envision a role for aesthetic and spiritual values in shaping social experience" at least partly "explained Dewey's acquiescence in the war effort." As an alternative, they developed "a highly subjectivist pragmatism that gave as much—and on occasion more—weight to the claims of love, intuition, art, and spirit in a critical philosophy of experience as it did to the deliberative reason and open communications that Dewey stressed."[29]

Another key concern of these personalist intellectuals was the increasing rootlessness of American society. Frank argued, for example, that when people lack a shared and embodied context where they *belong*, they are reduced to isolated atoms "moving alone in a herd through a bewilderment of motions."[30] The Young Americans believed that "beloved communities" could only be created in contexts where participants shared a deep sense of place and collective history. Prefiguring Dewey's celebration of the local at the end of *The Public and Its Problems* in 1927, they argued that an authentic common dialogue required a shared cultural and historical "frame" that would give context and coherent meaning to the individual contributions of each participant, allowing the development of rich common efforts. Like Dewey in *Public*, they struggled to conceptualize how this might occur in the modern world. Unlike Dewey, however, what was at best a secondary issue for him was a central problem for the Young Americans.

The Young Americans spent extensive time trying to figure out how one could recover or recreate what they sometimes called a "usable past" in an industrial world in which the deep common histories of premodern communities had largely been destroyed. They generally agreed with Dewey that the kind of rich interpersonal dialogue they sought could only take place within face-to-face communities. As a result, they generally envisioned cultural renewal taking place through a decentralization of American society. Myriad small communities would need to work to recover and create their own local histories and shared values and projects. How exactly this could happen remained as unclear in personalist writings as it did in Dewey's, however.

In their most cogent moments, as the discussion that follows indicates, the Young Americans understood that their disagreements with Dewey represented differences of perspective on a mostly shared set of commitments instead of completely different paradigms. In a simple sense, Dewey foregrounded the importance of democratic practice and tended to assume that individuality would emerge as a byproduct of truly collaborative contexts. The personalists, in contrast, stressed the importance of actively nurturing individual expression and authentic engagement between individuals, often assuming, for their part, that some kind of egalitarian democracy would naturally emerge as a result. One side stressed democratic technique and institutions while the other worried more about cultural and aesthetic renewal. But both valued individuality, democracy, and authentic, egalitarian interactions based on experiences in face-to-face communities of dialogue. Despite this common framework, however, their divergent perspectives cannot easily be integrated. As I argue later, it seems likely that there is, in fact, a fundamental tension between a focus on collaboration and on actualizing individual uniqueness, a tension that Dewey rarely acknowledged.

John Dewey versus Lewis Mumford

The ferment of the time and the criticisms of the personalists likely influenced Dewey's turn to questions of aesthetics in the 1920s and 1930s. Dewey's writings reflected little explicit acknowledgment of this influence, however. He rarely cited the writings of key personalists, and when he did he used the occasion largely to criticize. Nonetheless, Thomas Dalton has argued that his writing was deeply affected by the attacks of the Young Americans, pointing out a number of instances where Dewey obliquely if not directly addressed their arguments. Dalton argued that Dewey "had been put on the defensive by Bourne and his fellow critics," which may have led him even to suffer "from a failure of nerve," as his "ideas and leadership [came] . . . under siege." Dewey was reacting during this time not only to criticisms over his support for the war but also against the wider cultural arguments of the personalists, referring disdainfully to the "Waldo Frank-Raldolph Bourne bunch," who he said were simply "precocity seekers."[31] As late as 1940, Dewey referred in a letter to the Young Americans as simply "a Mutual Admiration Society."[32]

Dewey's most explicit engagement with personalist ideas came in the form of a response to Mumford's critique of pragmatism in *The Golden Day* (1926).[33] In a chapter called "The Pragmatic Acquiescence," Mumford accused the pragmatists—singling out Dewey and William James—of failing to challenge the growing cultural desert of industrial society. In Mumford's view, pragmatism had simply played into the spiritual emptiness of modern America, promoting the instrumentalism of industrial society and lacking any real sense of beauty or mystical completion. Like other personalists, Mumford believed that "true" art, requiring a deep engagement from viewers, listeners, and readers, had been replaced by the shallow entertainments of modern consumer society. (Think *Masterpiece Theater* vs. *American Idol*.) While Mumford acknowledged that Dewey was at least beginning to recognize "the place of the humane arts," Dewey rarely addressed this issue, focusing nearly all his energies on "science and technology." As a result, Mumford complained that Dewey's writings ended up serving society's obsession with "instrumentalism in the narrow sense, the sense in which it occurs to Mr. Babbitt and to all his followers who practice so assiduously the mechanical ritual of American life." Whether intentional or not, "what Mr. Dewey has done in part has been to bolster up and confirm by philosophic statement tendencies which are already strong and well-established in American life, whereas he has been apathetic or diffident about things which must still be introduced into our scheme of things if it is to become thoroughly humane and significant." In Mumford's view, even Dewey's celebration of democracy was simply part of this "acquiescence," embodying the belief that "what had been produced by the mass of men must somehow be right." Mumford's argument was very much a situated, historical one. It was partly the challenge of their *particular* historical moment that made Dewey's ideas so problematic.[34]

At the end of this chapter, Mumford turned to quotations from Bourne's angry critique of Dewey, written in the wake of Dewey's support for American entry into World War I. Bourne had lashed out at Dewey for destroying what Bourne saw as the real potential of pragmatic thought. "It never occurred [to us]," Bourne declared, that pragmatism would allow "values" to "be subordinated to technique." Missing the fact that Bourne was actually arguing that Dewey's action was really a betrayal of the true potential of pragmatism, Mumford concluded that pragmatism in general lacked "values that arise out of vision." As a result, he believed that Dewey ended up simply feeding the growing tendency for people to subordinate "their imagination to their interest in practical arrangements and expediences," canalizing "the imagination itself into the practical channels of invention" and leading to "the maceration of human purposes."[35]

Despite the cogency of much of Mumford's overall critique, it should be clear even from this brief summary that his interpretation of pragmatism contained inaccuracies and exaggerations. In a testy response published the next year in the *New Republic*, Dewey tried to correct Mumford, noting that values *were* in fact central to pragmatic thought. Instead of "acquiescing," Dewey noted, with perhaps some detectable weariness at explaining himself yet again, that pragmatism actually sought to bring values *back* into "science." Within the pragmatic vision, values became hypotheses—about the good life, about how to treat others—to be continually tested and evolved like everything else in the context of everyday experience. Similarly, he reiterated that his vision of democracy did not simply represent "acquiescence" to the aggregated opinion of the masses. Pragmatic democracy aimed not for unsupported opinion but for findings that emerged from joint efforts to intelligently test different approaches and solutions. And while the text of *Experience and Nature* has become famous for its obscurity, Dewey rightly noted that Mumford had misread its argument about art, which was actually quite similar to the one Mumford was making. Dewey understood that art was not merely "instrumental." He agreed that true art embodied social values, at its best "perfecting the potentialities of any and all experience."[36] While Dewey used his brief response mostly to point out Mumford's misunderstandings, at the end he went further, concluding that by rejecting the "instrumental" aspects of Deweyan pragmatism, Mumford had shown himself to be simply a utopian dreamer without the tools to actually achieve what he dreamed about.

In his reply to Dewey's criticisms, also published in the *New Republic*, Mumford began by noting that he was "honored that Mr. Dewey chose *The Golden Day* to stand the brunt of his attack." He pointed out, however, that *The Golden Day* was only the most recent iteration of an ongoing series of criticisms that he, Bourne, Brooks, and Frank had been making for at least a decade. "Where on earth," Mumford wondered, "has he been these last ten years not to have felt the sting of this criticism before?" Portraying himself as a representative of this group, Mumford used his reply to try to clarify what he and his colleagues thought was missing from Dewey's overall vision.[37]

Mumford began by emphasizing how indebted he and his colleagues were to pragmatism, declaring that he still saw himself as a "pragmatist" in many important respects. He noted, however, that since World War I pragmatism "has come to seem to many of us not false but insufficient." And he stressed that the differences between Dewey and the Young Americans were ones "of practical emphasis" rather than absolute disagreement. Mumford agreed that science and technology were, of course, important. But Dewey seemed to think that they were sufficient. It was "this lack in the body of Mr. Dewey's thought if not in its abstract outline"—the failure to emphasize the importance of aesthetics, especially—that had led Mumford's generation to reach "out to other thinkers," looking for those "elements that physical science and technology squeezed out of the foreground."[38] He emphasized that he and other personalists had not "lost contact with the industrial world or retreated from it" into a nether world of pointless utopian thinking. But the Young Americans believed that the modern world spent too much time obsessing about science and technique. As a result they looked elsewhere for arenas that seemed to have been lost in the meaningless hubbub of modern industrial society.

What was needed, Mumford argued, was not *only* the "scientist who says 'It must'" but *also* "the ability to think creatively with the artist who says 'I will.'" Again emphasizing the historicity of his argument, he stated that the central problem with Dewey's focus on science and technique, from his perspective, was that it had not sufficiently responded to "the desiccation and sterilization of the imaginative life" of their time. Mumford pointed to his own work on architecture as an example of a more effective balance between science and art, since architects think "both scientifically, in terms of means, and imaginatively, in terms of the humanly desirable ends for which these means exist."[39]

In his conclusion, he emphasized that "it is not that we reject Mr. Dewey . . . but that we seek for a broader field and a less provincial interpretation of Life and Nature than he has given us." In fact, while Mumford maintained his conviction that Dewey held a quite limited understanding of art and aesthetics, he acknowledged that Dewey seemed to be moving in their direction to some extent recently, at least beginning to embrace "certain aspects of modern art." Despite his discomfort with the tone of Dewey's response, Mumford was heartened to see that he was "still thinking experimentally and freshly, . . . reaching out to wider sources of experience."[40]

Mumford's reply seems to have made little impact on Dewey. In fact, three years later in 1930, Dewey extended on his brief critique of *The Golden Day* in *Individualism, Old and New*. This time he looked to Frank's *The Rediscovery of America* as an example of what was problematic with the personalist perspective of the Young Americans.[41] The fact that this was the only book cited by name in all of *Individualism* and one of only a handful of citations in the entire volume, indicated Dewey's special concern for the book's arguments. What Frank did not understand, Dewey argued, was that in an industrial society "literary persons and academic thinkers are . . . effects, not causes" of social change. "A sense of fact and a sense of humor," he noted dismissively, "forbids the acceptance of any . . . belief" that artists have the power to produce significant changes in society. In fact, reiterating his earlier attack on Mumford, he accused Frank of "indulgence in . . . fantasy." Art could become transformative in a significant way only when it emerged from changed social conditions. Along with the other conceptions of individuality that he critiqued in this book, what he argued personalists did not understand was that only the emergence of a truly democratic society would create conditions that could allow individuality and expression to flourish in any coherent and effective manner.[42]

As Dalton and this discussion have shown, Dewey was paying attention to the emergence of personalist thought in the 1920s and took the time to react to their arguments. At the same time, what we have available to us seem more like offhand comments than focused, analytical critiques. While the writings of these personalists and the personalist tenor of the times may have generally fed his increasing interest in aesthetics, then, it is unclear how the specific arguments of the Young Americans about the place of aesthetic expression affected him.

This is not particularly surprising. As the exchange with Mumford indicated, the Young Americans could be quite caustic, their writing often exaggerated and sometimes inaccurate in their rejection of Dewey's vision—especially in the case of Brooks and Mumford. Further, it must have been difficult for Dewey to separate critiques like Mumford's from the Young Americans' criticisms of Dewey and other collaborative progressives' support for America's entry into World War I—even if Dewey himself began to move in their direction after the war. This is especially true given Bourne's aggressive, dispirited, and, as has been widely noted, painfully accurate accusations that unfaithful pragmatists like Dewey had fallen to some extent into the darkness of valueless instrumentalism.[43]

Perhaps most important, the personalists never really settled on any single overarching theoretical framework. The mixture of insight and hyperbole visible in *The Golden Day*, combined with frequently shifting arguments, wavering commitments to democracy, and more, was reflective of the writings of the Young Americans more broadly. Struggling to find some adequate answer to what they saw as the cultural banality and herd-like conformity of modern America, they tried out many different possible approaches. It would have been difficult even for a careful reader to know exactly what to criticize.

Dewey's pattern of response to personalist thought in the 1920s was mirrored in his comments on personalist versions of progressive education that were also emerging during these years. His attacks on the "progressive" educational efforts of the time were part and parcel of his broader concerns about the "aesthetic turn" of the decade.

Key Defenders of Personalist Education in the 1920s

Progressive education during the 1920s reflected the emerging ideas of the broader collection of personalist intellectuals. While there was great diversity in the alternative approaches of different experimental schools, scholars have generally agreed, as Harold Rugg and Ann Shumaker argued in their widely read survey of the schools of that time, that a key characteristic was an effort to foster "creative self-expression" and the kind of organic, egalitarian community that Bourne, Brooks, Frank, Mumford, and others celebrated.[44]

Following the lead of earlier scholars (especially Robert Beck and Lawrence Cremin), I focus here on the two key progressive educators that seem to best reflect the pedagogical thinking of the 1920s: Margaret Naumburg with her Walden School and Caroline Pratt with her City and Country (or Play) School. In general, reports about progressive schooling published during those years, especially Rugg and Shumaker's *The Child-Centered School*, published in 1928,[45] indicate that Naumburg and Pratt's visions were fairly representative of most progressive schools of the era, even if other educators were not usually as sophisticated. Both lived and taught in Greenwich Village, the bohemian hotbed of personalist thought in the United States[46]—downtown from Dewey's office at Columbia—participating in the vigorous dialogues about these ideas in cafes and salons. Naumburg was actually married for some years to Frank, and they worked together on educational issues.[47]

The work of each educator reflected the influence of one of the two core theoretical streams of the era. Naumburg's practice was most influenced by Freudian psychoanalysis, especially Jung, while Pratt was more affected by theories of artistic expressionism. I do not emphasize these differences here because the fact is that these two perspectives were almost inseparably intertwined in the thinking of the time and because, partly as a result, the overall approaches to pedagogy developed by the two women were quite similar. This was especially visible in the focus in their schools on different forms of art as opportunities for individual expression and collaborative engagement, as well as in Naumburg's later interest in art therapy.

Naumburg and Pratt are especially relevant to a discussion of Dewey's critical response to personalist education because we know that Dewey was acquainted with both of them, although it is unclear how much experience he had with their actual schools. The discussion that follows focuses on Naumburg, making more general references to Pratt. Not only had Naumburg been one of Dewey's graduate students, she was also one of the few personalist educators willing to directly criticize Dewey in her writings. Beck even suggested that Dewey's critical essays on personalist education in 1929 and 1930 were written in direct response to Naumburg's criticisms.[48] Naumburg also may have been more deeply involved in the broader personalist intellectual dialogues of the time than Pratt, indicated by her marriage to Frank.[49]

Reflecting the concerns of other personalist educators of the time, Naumburg and Pratt feared that Americans "were losing their distinctively individual personalities."[50] And they developed pedagogies designed to help their students overcome the individual-destroying nature of modern bureaucratic society. While they faced "constant criticism[s]" from Deweyan collaborative progressives throughout the decade that "their philosophy of education left the students adrift in a miasma of freedom," they "held firmly to their belief that creative self-expression was the power most regrettably missing from the American scene."[51]

Naumburg and Pratt repeatedly emphasized that their schools did not simply give students total license to do what they wished, and both bristled at the accusation that teachers in their schools failed to "teach" at all. In their writings they showed that children in their schools were not simply left to their own devices to develop creatively. Echoing Nel Noddings's later writings on care theory, for example, one of Pratt's teachers noted, "You have to feel the thing the child wants to do, to think his thoughts, in short, to become a child yourself. And to be able to do so you must have the soul of a child and unless it is so, and only so, you can't get results."[52] They collected extensive observational data over time in their schools to show how their teachers used the environment and, when necessary, direct engagement to encourage students to engage in and extend on their work.

They acknowledged that their schools did not have and, in fact, resisted creating preplanned curriculum for their children. Nonetheless, they described clear patterns in the kinds of activities students of different ages were encouraged to engage in. Teachers in each school also clearly drew from a shared sense of what they were trying to accomplish, with similar approaches, for example, to asking questions to foster children's creative explorations of the workings of the world around them. And there was a set of common items in classrooms that gave children in specific age groups particular kinds of opportunities for play and exploration, including sets of specially shaped blocks designed by Pratt to encourage creative student constructions. Many of their activities were rooted in explorations of their local environment—visits to shipyards, printing plants, and local parks—followed by extensive dialogues prompted by teacher questions about their experiences.

Naumburg provided a good example of the pedagogy in her Walden School at the end of her book, *The Child and the World*, in a transcript of an interaction between a teacher and some students as they worked on creating a play about immigrants to New York. I have chosen a section where the teacher does more speaking to give a sense of her interaction style:

> TEACHER. Right now we are interested in getting the characters of these people. If we finally decide upon this scene and the father puts the hatchet on the chair, will she sit on it?
> EDITH. Oh, of course not.
> TEACHER. There is another new point that's very good that Lillian brought in. A new idea there. The greatest sorrow is not only money, but the idea of the loss of his children.
> LILLIAN. He wants to make his children happy.
> TEACHER. What do you mean—by the loss of his children?
> ALICE. Not that they're lost on the street. But they have gone from him.
> LILLIAN. Let's bring in the daughter.
> TEACHER. You know, the first person to come in need not be the father. It might be a neighbor. She [his wife] would be expecting him any moment, and somebody else comes in who's not the father. She might talk to the neighbor for a while and go on peeling potatoes.
> LILLIAN. She'd better be a man, so the wife can mistake him for her husband.
> ELSA. I want to be a neighbor.
> TEACHER. All right.
> ELSA. [walking in very calmly] Good day! Are you busy today?
> ALICE. I am very busy for my family.
> LILLIAN. She acts like an old woman [meaning Alice]. It's good.

ALICE. Maybe I would act differently in New York. Maybe, because of the bustle of New York, my voice would get much louder . . . [Jane takes the part of the neighbor.] Good morning.
JANE. [walking by] Good morning . . .
TEACHER. I didn't get any particular meaning out of Jane's part.[53]

Naumburg described this as an effort to "train . . . children to improvise." The following section is a good example of her explanation of this idea:

> We begin by spending a great deal of time in a class discussion of the material, whatever it happens to be, whether a traditional story or an original theme that they [students] have decided to dramatize . . . During this formation period of a play, the teacher's part is very important. For often fresh and original suggestions are made which an alert and creative teacher may be able to value more truly than the mass of the class. And just as she can often lead the group to reconsider a positive suggestion that they might have set aside, she can also steer them away from repetition of their old plots and stereotyped scenes and give them the courage to experiment freely for themselves. We all tend to drop back so regularly into old grooves that it is particularly necessary to help the children break away from the repetitions of our empty theatrical forms . . .
>
> In order to get the original plot worked into its final subdivisions . . . the outstanding episodes must be agreed upon. Again the teacher, by leading questions, can help focus the contending forces in a group.[54]

These citations give a sense of the richness of the kind of pedagogy Naumburg was promoting. Quotations from Pratt's writings would look quite similar. Note the frequent but often subtle teacher interventions throughout this process. The teacher is, in fact "teaching" improvisation, not simply letting the students do what they want and ignoring them. In fact, one is reminded of the kind of interventions the Laboratory School teachers used to nurture their students' thinking. The teacher sometimes stops children and creates "problems" to solve by asking questions and by creating situations in which their peers will raise issues that they need to address.

In both Naumburg's and Pratt's schools, the teachers worked hard to foster a healthy community. Children in these schools were given many opportunities to interact and work with other children in a safe and protected environment, supervised by teachers. They discussed, for example, their efforts to encourage students to balance their own desires with those of others and to work together on common projects. And they provided extensive empirical evidence that their efforts effectively nurtured respectful environments and collaborative efforts.

This evidence seemed to be largely ignored by critics, however. As Naumburg complained, they "imagine that we neglect the social aspect of education altogether." But "the truth is that we trust the innate social instincts of children so profoundly that we find it unnecessary to force socialization from above."[55] She argued that they should be judged by their *results* and not by misunderstandings of their pedagogy.

As was already clear in the previous excerpts, Naumburg's descriptions of the approach to socialization used in her school at times actually seemed to echo aspects of the social approaches of Dewey's Laboratory School. She noted, for example, that "in the right nursery-group environment . . . , normally self-seeking [impulsive activity], breaks against a world of similar entities [e.g., other students] . . . If the conditions of group-life are consciously planned and the teachers know just how to hold back or redirect the energies of the youngsters, it is possible for children as young as two or three to be thrust into conditions of daily life where the need and desire for playful co-operation creates the beginning of an organic group."[56] Like most of the more sophisticated progressive schools of the time, as described by Rugg and Shumaker, then, Naumburg and Pratt provided a range of opportunities for children to work together and a context in which they could learn to adjust themselves to each other's needs and collaborate on common projects. These activities included creating musical and dramatic performances and running a school store, a post office, a toy-making factory for younger children in the school, and the like.

Reports about leading personalist schools of the time generally praised their social environments, supporting these educators' contentions about their effectiveness. For example, Alvin Johnson noted that what Naumburg's "Walden School attempts to do, and in remarkable measure succeeds in doing is to create conditions in which each child, following out his own natural interests, spontaneously associates himself with fellow pupils and teachers . . . Those who have been trained according to Miss Naumburg's scheme will be more disposed to be engaged in active cooperation with their fellows than to hold aloof from common enterprise while paying the price of superficial conformity."[57] Note again, however, that the personalists framed cooperation much more vaguely than Dewey, focusing less on accomplishing common projects than on the development of a web of organic intimate relationships that fed the development of individual distinctiveness.

Given the wide range of personalist schools during the 1920s, there were surely some, probably many, that did not make this effort to intentionally foster healthy interaction. However, in *The Child-Centered School*, Rugg and Shumaker, reviewing evidence collected from more than one hundred different schools, reported that, in general, "through a variety of group activities . . .

the new school . . . sets up situations which provide constant practice in cooperative living. It encourages activities in which he can make a personal contribution to group enterprises; in which he has social experiences, graded to fit his level of social development; in which he feels himself an accepted and respected member of a society of which he himself approves."[58]

Like Dewey, Naumburg and Pratt believed that their pedagogies could contribute to the emergence of a better society. They hoped to encourage society to be more open to the unique vision of individuals. Pratt even "considered herself a radical socialist" believing that her efforts to develop a pedagogy designed around "play" would lead toward the development of a socialist society.[59] Along with other personalist intellectuals, these personalist educators believed that truly integrated, self-actualizing, and uniquely expressive individuals were the real foundations of a truly democratic society.

As with the work of broader personalist intellectuals, however, in the writings of personalist educators it is often difficult to discern *exactly* what the role of the teacher was supposed to be or the form the social environment was supposed to take. While they did provide evidence of more directive engagement and guidance than critics wanted to acknowledge, sometimes it did seem as if they were encouraging teachers to mostly leave children alone. Their ambiguity in this area may have resulted at least in part from the fact that they approached teaching more as an "art" than a "science." But it contributed to misunderstandings about their vision in the wider public realm and likely to misappropriations by other educators.

Finally, it is important to emphasize that as students grew older, personalist schooling often started looking more traditional. In the high school years, these educators often turned to a focus on more standard teaching and subjects. While the middle-class parents of these children might have accepted a "free" education early on, they nonetheless expected their children to be prepared for college. In other words, the personalists did not discard disciplinary knowledge, although they placed it much later in a child's education than Dewey or other traditional educators would recommend. The free schoolers of the 1960s, discussed in the next chapter, often did much the same. (Interestingly, in his own short-lived Laboratory School, Dewey and the teachers failed to solve the problem of how to effectively engage older children in his constructivist model.)

Naumburg's Critique of Deweyan Education

As I noted previously, in 1915 Dewey and his daughter Evelyn embraced the first glimmerings of the new progressive school movement as exemplified by Naumburg and Pratt. A little over a decade later, however, Dewey began to make more critical statements about current forms of progressive education. In 1928, in a talk to the Progressive Education Association he warned, as he would later on, that a simple reaction against traditional, authoritarian forms of schooling was not enough to ground a new approach to educational practice. Dewey was apparently still hopeful at this point, however, that this "more negative phase of progressive education," which involved simply removing "artificial and benumbing restrictions" might soon "run its course." He hoped that in the future the "new" schools might shift to "a more constructively organized" approach, in other words, the one *he* recommended. As part of an effort to contribute to this hoped for shift, he lectured his audience on the characteristics of *real* individuality. Distinctive, socially healthy individuality could, he asserted, only emerge within the kind of organized approach to content that he had long supported, in which children learned to work on long-term projects, connecting smaller efforts together in sequence and developing increasing knowledge and accomplishment over time. A "sequence of unrelated activities" of the kind he thought dominated child-centered schools would not, he noted, "provide for the development of a coherent and integrated self." And while he agreed that one must begin with the interest of a child, he stressed that without a teacher's assistance the child's activities were "not likely to lead to anything significant or fruitful."[60]

Leading members of the association like Naumburg and Pratt were likely insulted by the underlying assumptions of Dewey's talk, which simply added to years of similar criticism.[61] In fact, later that same year Naumburg published *The Child and the World* (its title almost certainly meant to contrast with Dewey's earlier *The Child and the Curriculum*), which responded somewhat angrily to the concerns of collaborative progressive educators and directly criticized Dewey's vision of democratic education. Fundamentally, she stressed her "strong dislike of Dewey's emphasis upon the social obligations of men in society rather . . . than their obligation to become individuals."[62]

Naumburg acknowledged that some interdependence between individuals was inevitable and important. But she argued that "interdependence ought ultimately to lead to greater differentiation and specialization of the parts, in the variation of individuals who make up the whole." In the end, then, interdependence "ought to lead *not to a sacrifice but to the true growth, of individuals.*"[63]

Dewey, she argued, ultimately misunderstood what society really needed from education. What the "world now needs," she wrote, are not people who are socialized to disregard those aspects of their uniqueness that do not contribute in some evident way to current social problems but "more complex individuals so that group life can become integrated on a higher plane." The problem with Dewey's vision of education, she concluded, was that he "places emphasis on the individual's living for the group rather than for himself." And she suggested that this focus on the group "is the cause of much of the world's suffering," since it is really "*the failure of the individual to know himself* [that] remains the [real] cause of his ineffectual adaptation to a positive and adequate social existence."[64] Echoing other personalist intellectuals, and again explicitly referring to Dewey, Naumburg went so far as to assert that "much of the present social philosophy that wishes to sacrifice the individual to the good of the group is *nothing but instinctive herd psychology*, translated into modern terms."[65]

Later in the book Naumburg extended on this critique of Dewey's vision. She reported on her own review of teacher records from the Laboratory School, which Dewey had made available to her, noting that she had found a "tendency to emphasize a uniform product in groups at a given period of organized work." She complained, "The making and doing of things was always subordinated to a social plan," which restricted any focus on "the individual capacities and tastes of the children," even "in the records of creative work."[66] As we saw in Chapter 2, she was right that the Laboratory School mapped out general curricula for different age groups around particular historical periods, with specific tasks that students generally engaged in (reinventing money as the Phoenicians or creating the bow and arrow as "primitive" people), although these remained responsive to the interests of the students. Naumburg also critiqued the Laboratory School's use of art and artistic activity, in particular, to serve the development of more systematic social and "scientific" understanding instead of as a strategy for nurturing individual aesthetic expression.

Despite the accuracy of many of her observations, however, Naumburg was overly harsh with some of these statements, echoing Mumford, Brooks, and other personalist intellectuals before her. And she likely understood that she was exaggerating. In part, her rhetoric may have revealed frustration with the difficulty of communicating what she was doing for years to the earlier generation of collaborative progressive educators. It should not be surprising, however, that language like this in Naumburg's book raised Dewey's hackles, almost certainly contributing to the harsh tone of his first published critique of personalist pedagogy a year later in 1929.

In fact, like Mumford, Naumburg's central argument was not that Dewey was completely wrong but that he had found the wrong *balance* between two extremes. Dewey did not understand that the personalist vision was, in fact, at least as social as his own—if more local and much less focused on common projects. Nor could Dewey see the danger that his narrow focus on collaborative activity might not make enough space for the development of unique individuality. A full flowering of unique expression, Naumburg believed, was not likely to emerge naturally in collaborative work. In fact, the Laboratory School's emphasis on contributions that supported shared projects did seem likely to suppress key aspects of individual expression that did not relate to these projects. As I understand her, Naumburg was arguing that Dewey's relentless emphasis on "democracy" and instrumental achievement in an industrial society was likely to become transformed, whether he liked it or not, into yet another example of herd-oriented collectivity. Only if educators included explicit efforts to nurturing individual expression were students likely to escape this fate.

Finally, as in the writings of the broader personalist intellectuals, it is important to acknowledge that Naumburg's intense focus on the enhancement of uniqueness was to some extent a *historical artifact*. The particularity of modern American industrial society at this point in history, with its increasingly bureaucratic tendencies, was part of what made her counterbalancing focus on uniqueness so critical in schools.

John Dewey Responds

In 1929, the year after the publication of *The Child in the World*, Dewey's guarded hopefulness about the personalist movement dissolved. The "new school" movement did not seem to be simply dying away as he had hoped in his address to the Progressive Education Association. Books like Naumburg's indicated that personalist educators were clinging to what he saw as their misunderstandings. His concern was not only with the inadequacy of the personalist approach. Even more problematically, some personalist pedagogues and their followers seemed to think what they were doing was actually reflective of Dewey's vision of democratic education. There was a great danger, then, that these spreading misinterpretations would end up distorting his own efforts.[67]

In an essay in an edited book on art and education with Albert Barnes, Dewey launched his first full-fledged attack on the personalists.[68] Extending on his 1928 progressive education talk, Dewey again accused personalist educators of lacking any coherent philosophical framework to guide their efforts. They were simply "reacting" against the "external imposition" of traditional schools by letting students do whatever they wanted.

"Such a method," he stated flatly, "is really stupid." Why? Because "it attempts the impossible," trying to educate students without actually guiding them in any way, "which is always stupid," misconceiving "the conditions of independent thinking." The real truth is that, "since the teacher has presumably a greater background of experience, there is the . . .

presumption of the right of a teacher to make suggestions as what to do." In fact, he noted, "the implication that the teacher is the one and only person who has no 'individuality' or 'freedom' to 'express' would be funny if it were not often so sad in its out-workings."[69]

In addition, what he saw as the failure to give students access to the accumulated knowledge of the disciplines was fundamentally counterproductive. Not only does providing students access to the "tradition" of particular arenas of knowledge not hinder the emergence of individuality, he argued, but such knowledge is actually what allows individuals to gain concrete power to achieve their own unique ends. "No one would seriously propose," he argued, for example, "that all future carpenters should be trained by actually starting with a clean sheet, wiping out everything that the past has discovered about mechanics, about tools and their uses, and so on. It would not be thought likely that this knowledge would 'cramp their style,' limit their individuality, etc."[70]

Here and in later essays, Dewey implied that he had significant experience of the workings and results of personalist pedagogy, although it was not clear exactly what experiences he was referring to. The only concrete example he used in his series of critiques was not of Naumburg's Walden School, Pratt's City and Country School, or any of a range of other leading schools. Instead he chose to discuss perhaps the most publicly reviled example of "free" pedagogy at the time: the artist Franz Cizek's approach to teaching art. Newspapers had widely reported and ridiculed Cizek's statements about how he taught art by letting children do whatever they wanted with no guidance at all. In other words, Dewey chose as his single example not a sophisticated setting like Naumburg's or Pratt's, but what could be considered one of the worst cases available at that time.

"Anyone who has seen Cizek's class," Dewey noted, "will testify to the wholesome air of cheerfulness, even of joy, which pervades the room—but gradually tend to become listless and finally bored, while there is an absence of cumulative, progressive development of power and of actual achievement." When such a misconceived approach fails in this way, he warned, the likelihood is that eventually "the pendulum" in the individual school and in the wider society will swing "back to regulation by the ideas, rules and orders of someone else, who being maturer, better informed and more experienced is supposed to know what should be done and how to do it."[71]

He returned to this argument in his next major salvo, the final essay in a series on the "new schools" in *New Republic* in 1930, in the issue directly following a contribution by Naumburg. He stated that his focus was on schools that seemed to carry "the thing they call freedom to the point of anarchy." But he mostly seemed to refer to the progressive movement in general. "Ultimately," he complained, referring again apparently to his personal experiences of these schools, "the absence of intellectual control through significant subject matter . . . stimulates the deplorable egotism, cockiness, impertinence and disregard for the rights of others apparently considered by some persons to be the inevitable accompaniment, if not the essence, of freedom."[72]

He did acknowledge the success of these schools in obtaining "for their pupils a degree of mental independence and power which stands them in good stead when they go to schools where formal methods prevail" and "in furthering 'creativeness' in the arts," an "achievement [that] is well worth while." These limited successes, however, were only "evidence of what might be done if the emphasis were put upon the rational freedom which is the fruit of objective knowledge and understanding." These achievements were "not enough." Ironically, given the argument of this book, he argued that personalist approaches were really only relevant to the middle class. While it "will do something to further the private appreciations of, say, the upper section of a middle class," he stated, "it will not serve to meet even the esthetic needs and defaultings of contemporary industrial society in its prevailing external expressions."[73]

In the years after these two key essays, Dewey continued to respond to personalist misunderstandings, especially seeking to refute simplistic visions of individuality. In *Individualism, Old and New* (1930), as we have seen, he criticized Waldo Frank.[74] In this and other writings, he reasserted his core arguments about the centrality of collaborative democratic democracy against the misguided visions of individual freedom he imputed to the personalists.

Even before he began to attack personalist ideas, however, he had started his own careful study of aesthetics, finally producing his magnum opus on the topic, *Art as Experience*, in 1934. In *Art as Experience* he examined how aesthetic engagement could generate the kind of integrated, unique individuality that Naumburg, Pratt, and other sophisticated personalist educators had long been emphasizing. Like the personalists, he drew deeply in this work from the romantic tradition in America, especially influenced by his own early encounters with European romanticism and American transcendentalists like Emerson.[75] Despite his public rejection of personalist pedagogy, then, his engagements with these currents of thought during the 1920s almost certainly helped drive his own explorations in this area. He never acknowledged this however. Nor did he ever really integrate these aesthetic ideas into his educational theory.

In fact, he used the occasion of his final small book (really a pamphlet) on education in 1938, *Experience and Education (E & E)*, as an opportunity to revisit his attacks on the personalists a decade prior. While his language in *E & E* was somewhat less blunt than before, his overall argument was nonetheless largely the same, beginning with his assertion that the personalists had no coherent philosophy of education. Again, he asserted, theirs was nothing more than a knee-jerk reaction against traditional schooling.

E & E extended on his earlier essays with the addition of a much more detailed critique of the personalists' failure to teach critical practices of scientific thinking and a wide-ranging critique of their failure to nurture collaborative democracy. Without access to sophisticated practices of collaborative, experimental inquiry, he argued, children in personalist schools would not develop skills for adapting their society to and guiding the rapidly emerging realities of the modern world. True freedom, he emphasized yet again, occurs not when children do whatever they wish, but only when they collaborate together, when they learn to coordinate their activities with others in service to common goals. Without such a focus, the unrestricted and unguided individual "freedom" of personalist pedagogy was actually "destructive of the shared cooperative activities which are the normal source of order,"[76] inevitably producing the "deplorable egotism" he had earlier asserted was rife in these schools. Free movement and free choice, Dewey stressed, are not ends to be desired but, instead, merely preconditions for the development of the sophisticated practices of collaborative democracy.

Overall, *E & E* expressed his ongoing fear that the wide dissemination of personalist perspectives would dash his hopes that a truly democratic form of schooling might emerge. By crushing personalism, he sought to leave the field open for his own vision of collaborative, scientific democracy.

Dewey's "Straw Person" Argument

As should be clear by now, in his critiques of personalist pedagogy Dewey consistently misrepresented what more sophisticated personalist educators were attempting. In Dewey's defense, while his own writings might often be dense or complex, at least they embodied a coherent perspective on effective democratic progressive education. The same was much less true of the personalist pedagogues, who, like broader personalist intellectuals, drew from a myriad of theoretical perspectives. Each drew from diverse interpretations of Freudian psychology, expressionist aesthetics, and other sources of romantic thought, as well as Dewey himself, in often idiosyncratic and sometimes not so careful ways.

Even individually these radical educators generally did not make the theoretical groundings of their particular approaches particularly clear. While Naumburg's explanation of her Jungian approach in *The Child and the World* was one of the most coherent, the form it took—a series of dialogues between imaginary participants—made it a challenge to tease out a single core perspective. And while Pratt made efforts to describe her aesthetic, expressionist approach in a series of publications, Mary Hauser has noted that she held a "strong aversion to specifying the nature of her philosophy."[77] Most "new school" educators were not as sophisticated as these two. In any case, Rugg and Shumaker found, personalist educators were often "enthusiastic rebels, fiery individualists, untiring explorers of child interests." Thus, "a detailed philosophical explanation of their method may have seemed antithetical to their overall vision."[78] More generally, these educators were not professional philosophers like Dewey. It should not be surprising, then, that their "philosophies" were more difficult to ascertain.

While Naumburg and Pratt did try to show in their writings how they provided opportunities for children to work together and adjust themselves to each other, for example, their descriptions of the path from individuality to democracy and of the *kind* of democratic practices this would entail were expressed in vague terms. Like other personalists, their writings often implied that the emergence of democratic communities and a democratic society would simply *happen* if the "integrated" individuality they envisioned could be achieved in local communities on a wide scale. Both the collaborative *and* the personalist progressives focused on the issues of most interest to them, failing to adequately address the challenges raised by the other camp.

These acknowledgments do not let Dewey "off the hook," however. The fact that the personalists often had difficulty or even resisted making their vision crystal clear did not relieve him of the responsibility of making sense of their efforts for himself. His failure to give serious consideration even to the work of Naumburg and Pratt and his apparent unwillingness to make serious attempts to elicit *any* coherent patterns of pedagogical thought from the range of materials available to him seem extremely problematic. A thinker of Dewey's vast capacity certainly could have done so if he had the inclination.

In fact, much of this work had already been done for him. He had surely read his Columbia colleagues' Rugg and Shumaker's book, *The Child-Centered School*. And other detailed reports about the pedagogy in key personalist schools were available as well. Despite some complaints, assessments of these schools were quite positive.[79] In fact, Rugg and Shumaker indicated that the limitations they did identify could be overcome without discarding core aspects of these schools' pedagogies. (Rugg himself remained a champion of personalist approaches.) They made it clear that children learned a great deal in these schools and also that children learned to live and to some extent collaborate well together in an egalitarian school community. Dewey's published criticisms did not acknowledge this broad body of supportive evidence, even though most of it was published prior to his first broadside in 1929.

Dewey's response to personalist pedagogues mirrored his response to personalist intellectuals more broadly. And the two were almost certainly closely linked. His resistance to personalist pedagogy was part and parcel of his resistance to the broad current of expressionist and Freudian theories during this era. His testy personal relationships with key members of the opposing camp, with Bourne and Naumburg and others in the intellectual hothouse that was New York City in the 1920s, probably contributed to his difficulty in taking their arguments seriously.

Overall, there is a sense in Dewey's push-backs against personalist educators that he believed he was addressing intellectual children. While he toned his language down in later critiques, even a decade later in *E & E* the implication remained clear that the personalists were, as he said in 1929, "really stupid" and not worthy of sophisticated philosophical examination. Dewey seemed unable to understand how the personalists could actually believe such astonishingly illogical ideas.

Ironically, with these "straw person" arguments Dewey was actually replicating, to some extent, the tactics of many of his own critics, then and later. By equating the failures of the most problematic personalist schools with the efforts of more sophisticated personalist pedagogues, he avoided the need to actually engage with the complexities of Naumburg, Pratt, and others. And this "guilt by association" approach was and would be in the future as effective against his own arguments as it was against the broader personalist movement.

It is true that some of Naumburg's statements—her assertion that his work was nothing more than "herd psychology," for example—seemed like unfair attacks on Dewey's overall vision. However, terms like these were in wide distribution among personalist intellectuals at that time, and she clearly assumed some familiarity with influential works by personalist cultural critics in New York during that decade (which, in fact, Dewey had). Her arguments needed to be placed in the context of this larger intellectual movement. Yet Dewey failed to make much significant effort to make sense of the contrasting perspectives of the personalists. Because he was unable or unwilling to "step into their shoes," as it were, he was unable to construct arguments—about educators or cultural critics more broadly—that engaged with their actual core commitments. Instead, he mostly focused on their failure to be good collaborative progressives in his own mold. Even then he missed the fact that many actually had read him quite closely and learned much from his writings.

Revisiting the Dewey versus Personalism Debate

Framed in less absolute terms, many aspects of Dewey's critique are well taken. The personalists did, in fact, focus much less on disciplinary learning than Dewey recommended. Their vision of the role of the teacher was much less clear. And, again, they had much less interest in initiating students into the kind of sophisticated practices of experimental examination or collaboration that Dewey championed. Further, Naumburg and Pratt focused their students' attention on their local environment, in the school and in the neighborhood beyond, in contrast with teachers in Dewey's Laboratory School who encouraged students to investigate the broad social, technological, and economic systems and relationships that held their entire society together. Education in personalist settings, then, was often much more parochial than in Dewey's. Dewey could easily have constructed a quite devastating critique of personalist education, at least from his perspective, even if he had taken their vision seriously as the embodiment of a coherent, contrasting philosophy of education.

If he had approached their work with a more open mind, however, he would also have needed to address the evidence indicating that students in these schools were not the "egotistic" monsters that he described and that, in fact, they ended up individually, at least, quite well prepared for engagement in adult life. He would have needed to distinguish more subtly between the more fluid and spontaneous yet still deeply egalitarian vision of community embraced by these schools and his own. He would have needed to examine the many collaborative projects personalist students engaged in and compare them in more detail with the kind of collaborative engagement he recommended himself. And he would have needed to grapple more sympathetically and respectfully with the broader arguments of the Young Americans and other contemporary personalist intellectuals whose ideas provided some of the most crucial foundations for their efforts.

Perhaps most important, an adequate engagement with personalist perspectives would have required him to take their criticisms of his own writings more seriously, especially their concerns about the limitations of his vision of collaborative democracy. As I noted previously, in their more cogent moments Mumford, Naumburg, and other personalists argued that Dewey's vision of democracy suffered from a lack of balance. In his effort to ensure that nearly all student activity consisted of coherent contributions to shared projects, they argued, he ran the risk of failing to fully nurture the unique perspectives of individual children.

While Dewey also deeply valued individuality, the actual practices he recommended did seem designed to encourage students to discard ideas and perspectives that could not be presented as constructive suggestions for a shared effort. In a discussion in 1922, for example, he argued that individual uniqueness is nurtured most "when the individual is working with others, where there is a common project, something of interest to them all, but where each has his own part."[80] In fact, in the Laboratory School almost anything that might lead individuals away from interdependence (like reading too early) seemed to have been removed from the pedagogy.[81] The implication of his work was that collaboration in which each group drew upon the capabilities of each person would nurture individuality as well. While evidence from the Laboratory School indicates that his approach achieved some success, Naumburg was almost certainly correct that there must be some trade-off between a focus on making coherent contributions to collaborative efforts and on encouraging the expression of unique perspectives within rich, intimate communities.

This trade-off was not necessarily a problem for Dewey in his earlier writings because he was most interested in those capacities that contributed to the improvement of the whole. It seems likely that he simply had not thought that much prior to the 1920s about aesthetic perspectives on individuality and expression, even though he understood quite clearly that the encouragement of unique individuality was critical to authentic collaboration and the development of a truly democratic society. As a result, he simply did not address the possibility that his focus on collaboration over free expression might shut down some aspects of individual development, closing off some avenues for the development of perspectives that might be socially productive in the future. While it was unfair to accuse Dewey of promoting the practices of the "herd," then, the personalists' more sophisticated critiques seemed close to the mark.

In general, despite Dewey's increasing focus on aesthetic practice during the 1920s and 1930s, including brief comments about the relationship between art and public action in *Art as Experience* and elsewhere, Dewey never really integrated his aesthetics into his democratic theory. Despite the high value he placed on individuality, as we have seen in earlier chapters, his focus on collaboration on shared projects meant that he never really spent any time on, and in fact sometimes seemed to deny the importance of (e.g., in Individualism, Old and New), nurturing aesthetic expression outside of such shared efforts.

In 1938, after the publication of *Art as Experience*, in response to the emergence of fascism in Europe, he did acknowledge that "I should now wish to emphasize more than I formerly did that individuals are the finally decisive factors of the nature and movement of associated life."[82] However, he went on after this admission to focus as he always did on the ways shared practices, processes, and institutions of democratic collaboration could foster individuality. It is not clear, then, exactly what the new implications of this admission were for his thought. Certainly he had not moved much in the direction of the personalists.

By providing children with extensive opportunities to practice a range of forms of expressive activities, especially aesthetic ones, personalist schools made extensive room for the actualization of such unique perspectives, placing few limits on the forms that these might take. The concerns of the personalists were broader than this, of course. What they were developing was a very different kind of educational context that, as a whole, left much more space for students to find their own routes to knowledge and understanding. As Naumburg noted, the personalist educators had more "trust" that with much less guidance than was provided in the Laboratory School, students would develop into creative, experimental, collaborative beings. However sketchily, personalist pedagogues in the 1910s and 1920s were trying in their writings to represent a broadly integrated vision of education, a vision from which no individual aspect could be eliminated without losing the whole that was much greater than the sum of the parts.

For this reason, it seems unlikely that one could simply integrate the insights of both sides together into a single pedagogy that captured the full benefits of both. Even personalists like Naumburg, with her fairly sophisticated understanding of Deweyan pedagogy, did not seem to believe that this was possible. Introducing a more directive and planned curriculum of the kind Dewey preferred, for example, would have done damage to the spontaneity and openness to student desires that the personalists deeply valued. Moving in one direction or the other seems to necessarily involve trade-offs between Deweyan or personalist commitments. The differences between these educational perspectives represent real tensions, requiring choices between what one will most value in the kinds of educational contexts one will create.

Conclusion: The Erasure of 1920s Personalist Pedagogy

As Robert Westbrook also concluded, a respectful and honest dialogue over the tensions between personalist and Deweyan democratic visions within and outside of education would have been of great value. Dewey's failure to take the time and to put forth the (emotional) energy to engage with the personalists in this way, therefore, was an enormous loss, the repercussions of which continue to reverberate today, especially through educational scholarship.

The arguments of 1920s personalists, much like those of the free schoolers of the 1960s and 1970s discussed in the next chapter, have almost completely faded from the collective consciousness of the field of education. Even the few examples of scholarly engagement with their perspectives have generally ended up downplaying the personalists' intellectual and practical achievements. Lawrence Cremin's influential but dated (1961) *Transformation of the School*, for example, presented a thoughtful overview of Pratt and Naumburg's work. But Cremin also implied that the "avant-garde pedagogues" of the 1920s simply did not read Dewey's writings carefully enough, ending his discussion by focusing, much like Dewey, on how the teachers generally did not "teach" in any significant way. Most treatments of Dewey's 1920s antagonists are less sympathetic than Cremin's. Among the few significant exceptions to this pattern are two essays on Pratt and Naumburg from the late 1950s by Beck in *Teachers College Record* and a brief recently published biography of Pratt.[83]

It is not much of an exaggeration to say that the almost complete elimination of the visions of people like Naumburg and Pratt from the education literature was a direct result of Dewey's dismissal of their work. Dewey decisively won the battle over the definition of progressive education. Regardless of the actual effect of his ideas on education in most schools, Dewey remains the preeminent philosopher of education in academia. As I have noted in earlier chapters, his perspective dominates writings in education about democratic education at the same time as his wider understanding of educational practice provides the grounding for nearly all branches of constructivist pedagogy. And because it is one of the best concise summaries of his educational views, it is *E & E*, out of all his works, that has become the most commonly used text for introducing educators and education scholars to his views. Those new to Dewey, then, often imbibe his critique of the personalists at the same time as they struggle to make sense of his educational vision. The intellectual bankruptcy of those I call the personalists, therefore, has essentially become an accepted "fact" among education scholars—to the extent they think about this issue at all.

From the perspective of this volume, one of the most problematic results of Dewey's demolition of the personalists is the way it has obscured the many similarities between his and their perspectives on education. To his deep chagrin, the personalists repeatedly stated that their efforts were deeply indebted to Dewey's own vision. I believe, however, that they were largely correct in many respects. In fact, I believe, as I have already noted, that Dewey and the personalists represented different positions on an essentially common spectrum of belief about human nature and learning. Their argument was more like an internal family squabble than a conflict between deeply opposed conceptions of democracy and democratic education.

This is difficult to establish with the evidence available about the 1920s, however, given the diversity of personalists' own explanations about exactly what they were arguing and why. As a result, I turn in the next chapter to a comparison of Dewey's educational vision with that of the personalist free schools movement of the 1960s and 1970s, the next and most expansive emergence of progressive pedagogy in the United States.

5

The Free Schools Movement

The previous chapter described how a group of young middle-class progressive intellectuals in the 1920s called for the reclamation and revitalization of the cultural wasteland of modern society. While retaining much that they saw as important from the "collaborative progressives," drawing especially from Dewey, they mapped out key limitations of the earlier generation's ideals. They sought to develop what I am calling a more "personalist" vision of progressive critique and social transformation. During the same years, a broad group of pedagogues founded new schools on a personalist model, sometimes in collaboration with these cultural critics. The work of these educators represented the first progressive education movement of any significant size to emerge in America.

Despite the real achievements of the radical thinkers and educators of the 1920s, and despite common themes, concerns, and pedagogical practices, an overarching theoretical framework never really coalesced. The Young Americans, for example, frequently shifted their core theoretical arguments as they cast about, mostly unsuccessfully, for solutions to an increasingly spiritually empty and "herd-like" modern world. The work of the personalist pedagogues of the time similarly embodied the great theoretical heterogeneity of the time.

Not until the 1960s, I argue, did a relatively clear, consistent personalist pedagogy develop in America that was grounded in shared sophisticated theoretical convictions. A loose collection of intellectuals—including Paul Goodman, A. S. Neill, and John Holt—helped develop a philosophical vision that prepared the way for and nurtured the second significant wave of progressive schooling in America: the free schools movement. These writers drew from many of the same romantic, aesthetic, and Freudian sources as had their 1920s forbearers, also looking to writings of the Young Americans themselves, including Mumford, who was still active. They were also influenced by new theorists, especially World War II European refugee thinkers in America like Herbert Marcuse. Because the work of the 1920s personalist pedagogues had essentially been forgotten by this time (even though some of their schools still existed), the educational ideas of Naumburg and Pratt, among others, seem to have had little direct influence. The relative coherence of the ideas developed by the 1960s personalists makes it possible to conduct a more sophisticated comparison with Dewey's educational theories than was possible in the previous chapter.

While the personalists of the 1920s and 1960s dealt with somewhat different issues and historical challenges, their overall visions of social change and education were quite similar. Both groups stressed the importance of enhancing individual distinctiveness within rich relational communities and sought to nurture children's capacities for creative responsiveness to their environment. But neither group tried to teach specific "scientific" or collaborative practices. And neither was much concerned about the specific "content" or disciplinary skills children ended up learning, at least prior to adolescence.

Interestingly, in contrast with the 1920s, few free school proponents were influenced in any direct way by Dewey. (Holt apparently never even read any of his work). Most personalist writers and educators were at least familiar with aspects of collaborative visions of progressivism, however. Some participated in Students for a Democratic Society (SDS), for example, and many had imbibed aspects of Dewey's ideas filtered through the contemporary writings of C. Wright Mills, among others.

As in the 1920s, the emergence of the free schools movement was partly driven by broader cultural trends. In the 1920s, however, young personalist intellectuals and pedagogues had formed a small, relatively isolated community amid the capitalist fever of a "boosterish" America. The late 1960s and early 1970s, in contrast, saw enormous numbers of alienated white, middle-class college-age students "dropping out" to join a vibrant personalist "counterculture." (For the sake of brevity, I will generally refer to this period simply as the 1960s.) And insurgent pedagogues among this group created new personalist schools in numbers much greater than in the 1920s. While the personalist schools of the 1920s represented scattered experiments, the explosion of personalist pedagogy during the 1960s was truly a "movement." Perhaps not surprisingly, however, most free schools seemed to lack particularly sophisticated understandings of the recommendations of the movement's key intellectuals. This, if you will, "free" approach to pedagogical practice—broadly understood—dismayed leading personalist educators like Neill, who followed up his breakthrough book in America, *Summerhill*, with *Freedom and Not License*, a book of sometimes testy responses to letters from those who seemed not to have read him very carefully.[1]

The 1960s and the Emergence of the Counterculture

In a broad historical sense, there were a range of interesting similarities between the eras of the 1920s and 1960s. These were, for example, both times of rebellion against the perceived stuffiness of an older generation, times when a broad collection of middle-class youth were seeking more individual freedom. And it is no accident that this happened during moments of economic security for the middle class. At both of these times, the key challenge for the new generation seemed to be individual and not group, cultural and not economic repression.

There were also important differences, however. Unlike the Young Americans, who came to prominence during the carnage of World War I, the revolutions of the 1960s took place a generation after World War II (the "good war"), in a culture also scarred by the Great Depression. And there were crucial differences in the core social concerns of key personalist thinkers. Perhaps most important, the Young Americans were still at least somewhat concerned about the place of the working class in America. The personalist thinkers who most informed the free schools movement, in contrast (despite the nearness of the civil rights movement), had much less interest in the "social question." They focused almost entirely on the challenges of their own class, trapped in the technocracy of the modern social machine. As James Farrell noted, afraid that they might be "potential victims of American culture," the people those who participated in the counterculture of the late 1960s "most hoped to preserve were themselves."[2]

In many ways it is useful to see the free schools movement as the pedagogical expression of the counterculture and its communitarian communes, growing and then fading along with the counterculture during the 1970s. Personalist intellectuals like Goodman fed the counterculture with their "encompassing criticism of the depersonalized institutions of mainstream culture." Participants "tuned in" to rock music, sex, and drugs and "dropped out" of bureaucratic society, creating a vast range of local communal experiments and hoping to "personalize America" through their example. "The counterculture," Farrell noted, "applied its personalist perspectives in a multitude of signs, symbols, institutions, and practices ... Suspicious of the instrumentalism of party politics, they practiced a 'magic politics' focused on individual transformation as the foundation of social change. Championing community in a culture of individualism, they tried to live in collaborative harmony with each other and with the earth ... Countercultural politics ... was the politics of anarchism—of decentralization, rural romanticism, and libertarianism."[3]

The Key Scholars of Personalist Education in the 1960s: Goodman and Neill

This chapter focuses on the writings of two of the most influential intellectual figures of the free schools movement: Paul Goodman and A. S. Neill. Goodman was seen by some as the "Father of the New Left," serving as "a guru to New Left social and educational reformers" and others during the 1960s. A controversial figure, Theodore Roszak described Goodman's "style" as one "that annoys into being taken seriously." Goodman was an eclectic independent scholar and artist—very much in the model of the Young Americans—who represented no particular academic discipline and whose skill lay not in his original insights but instead in his ability to draw a range of ideas and material together into an accessible form. Although a prolific writer in many genres for much of his adult life, it was *Growing Up Absurd*, published in 1960, that largely made his name. In this book an entire generation of white Northern college students found their often vague discomforts and complaints about modern society given coherent voice. *Growing Up Absurd* was, for many, the "'bible' of the New Left."[4] In part an attack on traditional educational institutions, the book provided free schoolers with key aspects of their philosophical foundation. Goodman also participated actively as the free schools vision coalesced into a movement—challenging, nurturing, and critiquing—until his untimely death in 1972.

Almost alone among prominent free school intellectuals, Goodman deeply respected Dewey.[5] In fact, like the Young Americans before him, Goodman argued that his own educational vision represented an extension and not a rejection of Deweyan pedagogy. Like 1920s personalists, however, Goodman felt that Dewey's vision had been superseded because of the emergence of new forms of oppression in modern society that Dewey had not foreseen. Goodman retained many essentially Deweyan commitments while, at the same time, transforming Dewey's "democratic" vision by combining it with a broad range of other work. Overall, Goodman's writings included an eclectic mix of Freudian psychological theories, aesthetic commitments, and anarchist political ideas.

Neill's writings about his pedagogy at the most famous of British free schools, Summerhill, were key influences on Goodman. By the time Neill's work found an audience in America, he had been developing his vision of "free" schooling in England for nearly a half century. For decades he had been Britain's most radical torchbearer of educational freedom.

Neill saw his role in very practical terms. In fact, when Goodman critiqued him for his apparent revulsion for theory in a letter, Neill happily acknowledged that he was not much interested in "theories."[6] While Neill was much more familiar with psychological theory than he usually let on, he generally disdained such "fancy" thinking. Instead, much like Pratt, he saw himself as an intuitive experimentalist, using his British boarding school, Summerhill, as a laboratory of "freedom."

In different works, Goodman held Summerhill up as one of the most advanced examples of the kind of "free" schooling he promoted. In fact, in *Growing Up Absurd*, Goodman declared that "the new progressive theory," after the obsolescence of Dewey's, was "'*Summerhill*.'" One of the few in the States who had read Neill's earlier writings, Goodman gave glowing references to Neill in *Absurd* that surely contributed to Neill's growing popularity. If *Absurd* was sometimes seen as the "'bible' of the New Left," more generally, Neill's *Summerhill* was often referred to as the "bible" of the free schools movement in particular. Like most of Neill's work, *Summerhill* consisted largely of vignettes about his school—tales of individual students, teachers, moments of struggle, and so on—that provided the basis of his arguments for a particular model of "free" education.

Partly because of Neill's iconoclastic style, it was easy to misread the educational argument in *Summerhill*, and many free schoolers did. It often seemed that Neill was arguing for just the kind of simplistic "false" progressive pedagogy that Dewey rejected. Neill's numerous assertions about children's inborn capacities for self-development fed this misconception. "A child is innately wise and realistic," he declared early on in *Summerhill*, noting that "if left to himself without adult suggestion of any kind, he will develop as far as he is capable of developing."[7] In fact, Neill did often argue that students should be left largely alone to do what they wished, released from any restriction of general rules. Perhaps most famously, children at Summerhill only went to class when they wanted to—some never attended formal classes for their entire time there. Moreover, surprisingly for an educator, Neill had no interest in sophisticated pedagogy. In fact, he worried about the dangers of sophisticated but potentially manipulative forms of progressive education like Dewey's. (As far as I can tell, Neill never engaged with Dewey's writings in any substantial way, declaring with some satisfaction that he had not even bothered to "go visiting" other progressive schools.)

Although he rejected efforts to assert control over students' academic learning, Neill was deeply concerned about the social development of his children. "If the emotions are free," Neill once said, "the intellect will look after itself," and this might usefully be taken as the central motto of his educational vision.[8] Thus, Neill rejected the focus of traditional educators on cognitive development. While Neill provided little recognizable guidance about pedagogy, then, *Summerhill* was filled with recommendations for often quite subtle interventions in children's social development.

While Goodman and Neill did not agree on every aspect of a "free" education, I nonetheless focus on their similarities—as most thoughtful free schoolers would have at the time. This chapter seeks to integrate the complementary aspects of their visions into a coherent and nuanced model of personalist pedagogy. Ultimately, Neill's was more an anarchist vision (focused on freedom) than a personalist vision (which added a focus on actively nurturing individual expression), although Goodman also considered himself an anarchist.

Human Nature: Comparing Dewey, Goodman, and Neill

At this point I turn to a more detailed examination of the comparisons between Dewey's vision and that of the free schoolers. I begin at the most basic level, describing Dewey's vision of "human nature" in more detail and showing how similar it was to Goodman's.

John Dewey: Impulses and Scientific Thinking

A central aspect of Dewey's argument against the personalists in *Experience and Education* (*E & E*) and elsewhere was that we should not expect children to learn advanced social and intellectual skills if they are simply left to their own devices. While Dewey acknowledged that children were perfectly capable of learning even quite complex practices—like language—in the absence of structured educational contexts, he feared that such unplanned learning usually only ended up reproducing the established society of adults. There was, he argued, little hope that children might *systematically* learn critical capacities for scientific investigation or collaborative skills of democratic engagement unless these were actively taught. Only if they were initiated into these practices, through approaches like those he and the teachers used in the Laboratory School, would they be able to creatively and "scientifically" reconstruct the society they were born into instead of simply accepting it as given. This was a central part of his attack on the 1920s personalists.

In other writings, however, and even at points in *E & E*, Dewey complicated this argument. In fact, from the beginning his educational model was based on the conviction that children are not simply passive receptors of information, that children are always actively engaged in making sense of the world around them. Despite these acknowledgments, however, Dewey could be vague in his educational writings, especially, about the underlying, presocial processes that drove this active engagement. Not until the publication of *Human Nature and Conduct* (*HNC*) in 1922, only a few years after his educational magnum opus, *Democracy and Education*, did Dewey comprehensively lay out his mature beliefs about the characteristics common to all human beings as biological organisms. And it is here that crucial similarities between Dewey's perspective and Goodman's and Neill's can be seen. Interestingly, as I noted in the last chapter, *HNC* was partly written in response to personalist and related Freudian visions of psychology in the 1910s and 1920s.

In *HNC*, Dewey distinguished between "habits," which, as described earlier, encompass all learned capacities for action, and "impulses," the relatively free energy that provides the motive power for all human activity, including habits. There is no action, he argued, without an impulse to propel it. Although impulses are innate, learning can alter the way they are expressed. While all sentient organisms from simple animals to human beings are driven by impulses, human impulses differ from animal ones in their flexibility. In most cases, animal impulses are permanently harnessed to relatively fixed biological instincts. In human beings, in contrast, almost "any impulse can become organized" through training "into almost any disposition" or habit.[9] This incredible adaptability of our impulses was, for Dewey, one of the key characteristics that makes us human.

The plasticity of our impulses allows humans to learn an almost infinite range of practices, but it also raises a set of challenges for human development since, by itself—before a human impulse has been harnessed to a habit—it "is as meaningless as a gust of wind on a mudpuddle."[10] Unlike many animals, human babies do not automatically develop capacities for survival and action without extensive and continuous support. Only when nurtured do we slowly learn to sublimate different aspects of our undirected impulsive energy into particular skills and abilities.

If our impulses did not have any inherent tendencies at all, however, it would be impossible to link them to particular habits. Like windup toys, we would simply dissipate our energy without ever learning anything. While impulses might initially seem entirely chaotic, then, they are not. In every case, Dewey argued, an impulse represents an effort by an organism to engage with its environment. Impulses can become harnessed to social practices because learned habits provide them with increasingly coherent opportunities to interact with and express themselves in the world around them. Babies, for example, do not simply cast about randomly. Instead, theirs is a ceaseless, often frustrating effort to respond to their shifting, disorganized experiences, a constant struggle to *make sense* of the world around them.

While the expression of impulses can be modified through social habits, their basic tendency to engage with and express themselves into the world cannot. The fundamental rules of impulse energy are as much a part of our essential biological inheritance as instincts are for animals. The constant emergence of impulses in human beings can never be fully halted or suppressed. Impulse energy, Dewey stressed, "is no more capable of being abolished than the forms [of energy] we recognize as physical." In fact, "if it is neither exploded or converted," it invariably turns "inwards, to lead a surreptitious, subterranean life."[11] Impulses run their course, one way or another, whether we want them to or not.

Because impulses never cease to engage with the specifics of the environment, it is never possible to turn people into happy robots who simply do what they are told. The inexorable pressure of impulses continually resists attempts to eliminate individual uniqueness and unpredictability. In fact, as Goodman would later examine in detail, Dewey acknowledged that efforts to suppress impulses can even lead to the emergence of "mental pathologies," something Dewey argued was established by "studies of psychiatrists [who] have made clear that impulses driven into pockets distill poison and produce festering sores."[12]

This "physics" of impulse energy helps explain Dewey's descriptions, written decades before *HNC*, of student responses to the regimentation common in traditional schools.[13] Because youth in these schools lack productive avenues for expressing their impulses, they often lose themselves in aimless daydreams and in "frivolous" activity, sometimes even resorting to drug use or other options that allow festering impulse energy to be dissipated. Even in the most authoritarian schools, the action of impulses produces a continual current of "underlife."[14]

While efforts to repress impulses are doomed to fail, Dewey believed that people subjected to repression frequently lose capacities for intelligent innovation. This is why members of traditional societies often cling to the way things have "always" been, even as the world inevitably changes around them. Instead of experiencing impulses as indicators of their own unique preferences and desires, as possibilities for understanding, regimented people treat them as internal enemies of their learned, "status quo" self. A tendency to destroy capacities for creative engagement is especially evident in authoritarian institutions like traditional schools.

The idea that children are always actively engaged in understanding their environments and that efforts to repress this activity can lead to pathologies was evident in many of Dewey's educational writings. Further, as early as his "Reflex Arc" article in 1896, it was clear he understood that learning was an essentially experimental process in young children.[15] In *HNC*, however, Dewey made his argument more explicit and linked it to his theory of impulses. In essence, he argued that *because of the natural operations of impulses*, children are *naturally experimental* in their interactions. Because the young are "not as yet subject to the full impact of established customs, their life of impulsive activity [remains] vivid, flexible, experimenting, curious." Capacities for experimental engagement are not created by culture, then; they are not socially acquired (although culture can refine them). Instead these tendencies exist from the beginning and remain vibrant unless culture damages them by preventing impulses from engaging with the specifics of a child's experience. For this reason, the relatively "unformed activities of childhood and youth" always—in every culture and context—contain "possibilities of a better life for the community." We are right then, Dewey argued, to "envy children" for "their love of new experiences . . . [and] their intentness in extracting the last drop of significance from each situation," as long, at least, as this love remains unblocked by the strictures of society.[16]

Of course without guidance the natural, fluid openness to possibility and experimental tendencies of children will never transform themselves into the sophisticated process of "scientific" action and reflection that Dewey cherished. As he emphasized in *E & E* and elsewhere, left to their own devices children will not recreate algebra, biology, or cooking practices. Without careful direction, children may learn only haphazardly, if at all, how things are connected to each other, especially in the modern world where these interactions have become increasingly distant from the day-to-day life of their local community.

On a more basic level, however, Dewey's acknowledgment in *HNC* of the naturally experimental tendencies of youth complicated and nuanced his dismissive critique of "false" progressives in *E & E*. If they are given space to develop in their own ways, *HNC* implied, one could imagine, at least, that children might retain (or regain) significant capacities for active learning and experimentation. And such capacities might at least begin to resemble those that Dewey and the teachers worked so hard to teach students in the Laboratory School.

Thus, you might be able to envision a form of "free" education that looks radically different than Dewey's that still nurtures the emergence of critical, creative thinkers. And this, I will argue, is what more sophisticated members of the free schools movement (and before them the personalists of the 1920s) did.

Goodman: Impulses in Gestalt Therapy

In the 1950s, in collaboration with Frederick Perls and Ralph Hefferline, Goodman developed a psychological theory that looked very similar to the vision Dewey presented in *HNC*. Although few free schoolers may have read the book they produced, *Gestalt Therapy*,[17] Goodman's later *Growing Up*

Absurd was in many ways a logical extension. In fact, Taylor Stoehr argued that *Gestalt Therapy* "was not superseded in Goodman's thought by New Left politics, but rather served as that politics' grounding in a theory of human nature and face-to-face community." *Gestalt Therapy* promoted a society of "ongoing group therapy," which became his model for education.[18] While Neill did not explore the nuances of human nature in such detail, he was deeply immersed in the emerging psychological theories of the day, and his practice at Summerhill largely agreed with the outlines of Goodman's vision. More generally, free school writers and practitioners shared convictions about what counted as "healthy" human beings and about what "healthy" pedagogy looked like that were grounded, explicitly or implicitly, in *Gestalt Therapy*'s basic vision of human nature.[19]

As Dewey had in *HNC*, in *Gestalt Therapy* Goodman and his colleagues argued that humans are driven to grapple with the world by the continual upwelling of what they called "emotions," or "impulses."[20] It was, they stated, the basic nature of human impulses to seek engagement with the unique complexities and contingencies of the environment. And impulse "energy" could not be eliminated or simply wished away. If it was not allowed to express itself somehow, it would necessarily build up inside people. While there were, of course, subtle differences between their specific understandings of impulses, then, *Gestalt Therapy*'s description of the function of impulses in humans was very close to Dewey's.

As in *HNC*, *Gestalt Therapy*'s model of human nature emphasized that all human beings, from their earliest moments, are active, experimental learners. Unless this ability is damaged by an inhospitable world, humans are to some extent capable of self-organizing and self-regulating themselves, of guiding their own development.

Oppression, Impulses, and Social Class

Of course, the actual educational practices championed by Dewey, and Goodman and Neill diverged in significant, often seemingly radical ways. Goodman argued that their divergence was the result of their very different conceptions of the nature of oppression in modern society. In other words, according to Goodman, personalists came up with different forms of education in part because they were designed to respond to very different social problems.

Dewey's Vision of Oppression

Dewey argued that oppression and inequality in modern and premodern times were quite similar. Across the ages, he argued, most people have been initiated into regimented social practices that do not allow much critical, creative thought. Those nearest the top of the social ladder, however, have been given more opportunity to think and plan, while those at the bottom have generally been perceived as closer to "animals"—less capable of complex thought. The main function in society of the lower classes has been to carry out the directives of those at the top. One could, he argued, see much the same pattern in the distinction between nobles and serfs in the past and in the distinction between factory workers and the engineers who told them what to do in his time.[21]

Goodman's Vision of Oppression

Following in the footsteps of the Young Americans, 1960s personalists like Goodman, in contrast, were often deeply, "romantically" nostalgic about premodern village societies. While Goodman understood that people in premodern contexts faced many social constraints, he believed that these had been relatively loose compared with the bureaucracy of modern society. Because social control in traditional societies operated mostly on a personal, face-to-face level, individuals had frequent opportunities to respond authentically to the unique aspects of particular situations and persons. In modernity, in contrast, individuals found themselves caught within a sophisticated social machine that allowed few avenues for individuality, creativity, rich relationships, or satisfying and productive labor. Increasingly, the remnants of the rich communities of traditional society were being destroyed by an "organized," all-encompassing "system."

Echoing Dewey's own earlier arguments in *The Public and Its Problems*, Goodman argued that modern society had "grown out of human scale," its interrelationships becoming enormously complex and distant from local experience.[22] For Goodman, however, the key problem with this new comprehensive system of modernity was new restrictions on individuality. The only way to succeed was increasingly to play the roles the system expected. For "the mass of our citizens in all classes," he argued, "there is no place for spontaneity, open sexuality, free spirit." Increasingly, people learned that any feelings falling outside their assigned roles should remain "private," and these unsanctioned feelings began to seem "freakish" even to themselves. Society seemed to have decided all possibilities "beforehand," preempting possibilities for creative change.[23]

Prior to the coming of modernity, Goodman explained, more people had "vocations"; they saw themselves as belonging to and contributing to something important in their communities. But modern work had increasingly become "determined not by the nature of the task but by the role, the rules, the status, and the salary; and all these" have become "what a man is." In the modern organization, people run "a rat race in a closed room."[24] They are oppressed, feel alienated, but have lost any way of framing this feeling of alienation. In such a world it increasingly becomes impossible to even imagine coherent alternatives to the status quo.

There was no one even to complain to about this situation. Earlier forms of bureaucracy, however restrictive, had always included channels of control, specific persons or groups one could identify that had the power to influence its actions. The technocratic "system," however, "exists only in the bland front of its brand-name products and advertising." As a result, "there is no knowing how it is run or who determines." Any "sense of initiative" and "causality" or of possibilities for individual or collective agency are destroyed.[25]

Schools were a key component of modern technocracy. In school, Goodman complained, "the mass of our citizens in all classes learn that life is inevitably routine, depersonalized, venally graded."[26] Schools only *seemed* more civilized with the elimination of corporal punishment and other overt forms of repression. As Michel Foucault and others would later also argue, "ideological exposure" in such institutions was becoming "unusually swamping, systematic, and thorough." As a result, Goodman argued, "authentic progressive education" must now move "in to new territory altogether."[27] Dewey's model of progressive education needed to be pragmatically evolved, as Dewey himself had often acknowledged, to meet new conditions that Goodman felt Dewey had not understood.

Goodman, Neuroses, and the Predicament of the Middle Class

Dewey's educational model was designed to eliminate the class-based intellectual distinctions and inequalities that he saw as the key forms of oppression in modern society. He wanted everyone to have the power to control their own working lives, and he sought a democracy in which everyone's voice would be heard and taken into account in planning the evolution of the larger society. By helping children understand that they could learn from anyone, that their community was always better if they opened space for others to participate equally in efforts to promote the common good, Dewey sought to teach those at the top and bottom that the system of hierarchy itself was destructive. In other words, a central goal of the Laboratory

School was to teach students, through their participation in joint efforts with others, to repudiate central aspects of the class system that gave some groups intellectual and material privilege.

The free school intellectuals' vision of education reflected their very different understanding of the oppression of modern society, focusing on the plight of the middle class. Unlike Dewey, who stressed *continuities* between processes of regimentation and social control in modern and premodern communities, Goodman and other personalists focused on *discontinuities*. In contrast with previous societies, which had depended on identifiable external authority for their continuing existence, modern society had come increasingly to depend on internalized self-control in its citizens. While citizens of premodern and modern society both tended to treat unused impulses as internal enemies, in modern society avenues for dissipating these "free" impulses had increasingly become closed off.

From a psychological standpoint, the key indicator of this new domination by *no one* was the increasing dominance of what the authors of *Gestalt Therapy* called "neuroses" in the personalities of modern citizens. As both Dewey and Goodman understood, impulse energy does not disappear unless it is expressed. In modern society, *Gestalt Therapy* argued, people constantly expended energy to keep their "foreign" impulses bottled up. When this suppression of a particular impulse becomes relatively permanent, when it becomes an essentially unconscious, automatic process it becomes what *Gestalt Therapy* called a "neurosis." Each "neurosis," then, is the result of an ongoing conflict with oneself.

The mere existence of neuroses was not necessarily a social problem, however, nor was the mechanism of the neurosis a new phenomenon. Throughout human history the emergence of neuroses had often represented a quite healthy response to dangerous contexts, preventing people from acting inappropriately. In modern technocracy, however, neuroses had increasingly become the central tool of human domination. In schools, in the media, in the family, and on the job, modern citizens learned to "choke" themselves off, preventing the expression of those parts of themselves that would allow them to engage authentically with their environment.[28]

Perhaps most problematically, modern neurotic individuals had became increasingly unaware of the ways they were inhibiting themselves. They "forgot" that they were the ones "doing the inhibiting" as the inhibition became "routine."[29] While modern citizens had a vague sense of incompleteness and of the insufficiency of their lives, they had lost any tools for identifying the problem and addressing it. Instead, they simply carried on, day to day, doing what society expected from them, numb to the richness of the world around them and to possibilities for acting differently. Their own personalities, their "souls" (as Foucault would later describe them) increasingly became neurotic prisons of their own making.[30]

While Dewey and Goodman shared core beliefs about the workings of impulses as a component of presocial human nature, then, they disagreed about how modern society distorted impulse expression. Dewey argued that the general failure across history to provide useful outlets for impulse energy forced individuals to find illicit and unproductive or explosive avenues for releasing the pressure that continually built up within them. Goodman, in contrast, argued that the prevalent form of oppression under modernity, the internal neurosis of blocked and *unexpressed* energy, was radically different from the past.

This divergence in their understandings of modernity, a difference that Goodman attributed to his belief that a technocratic society began to emerge with real strength only near the end of Dewey's writing career, led to very different evaluations of the workings of social class. Dewey argued that upper- and middle-class members of society have, throughout history, generally been allowed much more freedom of thought and action than those on the bottom. With his theory of neurosis, Goodman essentially reversed this gradient of domination. While Dewey's vision might have been accurate in the past, under technocracy Goodman believed that social "success" was increasingly tied up with subjection to intensified systems of *neurotic* control. While those on the bottom might experience more material want, the people higher up, Goodman and other key 1960s intellectuals argued, were most psychically oppressed.

In *Growing Up Absurd* and elsewhere, Goodman was usually careful not to romanticize the position of the poor and the working class in modernity. He acknowledged that "a major pressing problem of our society is the defective structure of the economy that advantages the upper middle-class and excludes the lower class."[31] From his perspective, those at the economic bottom of society faced a tragic choice: either they remained poor and marginal to the larger society but more psychologically free, or they subjected themselves more fully to the "system" through advancement into the middle class.[32]

Still, the personalists' general sense that those on the bottom were more "free" than those "above" them could lead occasionally in Goodman's writings and that of other contemporary personalists like Edgar Friedenberg to a nostalgia for economic marginality and material oppression. In contrast with numb neurotics who were fully incorporated into the technocratic system, those outside the system could at least experience some freedom of impulse expression. At one point, Goodman himself even argued that "at present in the United States, students—middle-class youth—are the major exploited class." His point was not that college students suffered more in any material sense, but instead that their "powers and time of life are used for other people's purposes" much more efficiently than were the powers of the working class, who experienced much cruder forms of control. In fact, he worried that students' material privilege actually "confuses them in their exploitation."[33] In other words, while the poor also suffer, the suffering of the middle classes and upper classes is much more insidious and difficult to address. (One can actually see in the writings of the 1960s personalists seeds of the fascination with similar postmodern visions of oppression that captured academia in the decades to come.[34])

This tendency to focus on the problems of the middle class was intensified by a belief common among relatively privileged free schools proponents that the elimination of material want was just around the corner, echoing similar beliefs in the 1920s. Poverty, from this perspective, was simply an artifact of technocracy that maintained unnecessary inequality amid abundance, forcing most people to work their lives away within the system even though there was no real need for such work. Because everyone was inexorably becoming middle class, it made sense that Goodman and other alienated interpreters of the 1960s saw the predicament of the middle class neurotic as the central political problem of their time.[35]

As usual, Neill did not expound these kinds of elaborate social theories. In fact, one biographer, Jonathan Croal, argued that "Neill's view of the world outside Summerhill was often a simplistic one." In a general sense, Neill agreed that the "school system 'had the political aim of keeping people down,' and of turning them in to wage slaves." And like American personalists, he was uninterested in "prosperity" or in efforts to make a living. In fact, he seemed more worried about the dangers of wealth than the dangers of poverty. To the extent he had a politics at all, it echoed the concerns of antitechnocrats like Goodman, embodying a fear of the deadening effect of success instead of the fear of material want. He even noted that he was disappointed that African Americans' "one idea was to get a white man's education."[36]

In any case, like that of the free schoolers, Neill's was an educational model largely developed with and for the children of the middle class. After a few early years working with Homer Lane in a school for "delinquents," he never made any effort to branch out beyond this group.[37] As a private boarding school, Summerhill could only take the children of families who could pay. There are even indications that Neill actually preferred children from these backgrounds. He noted, apparently without irony, for example, that "our successes are always those whose homes were *good*."[38] Summerhill seemed to best serve children who arrived imbued with a particular culture, a particular form of "cultural capital."[39] And in Britain, just as in the United States, "he reached few parents of the working classes."[40]

The Ideal Individual: Deliberative Scientist or Performance Artist?

As a result of their different understandings of the challenges of modern society, Dewey and the free schoolers developed very different perspectives on the kinds of *persons* they hoped education might foster. Dewey's ideal was that of the deliberative, everyday "scientist," continually engaged in efforts to understand the workings and relationships of the world around him or her. For Goodman and Neill, in contrast, the ideal person was something like a "performance artist," dancing fluidly in response to the complexities and contingencies he or she encounters in the material and social world. Again, however, these ideals are not as different from each other as they might appear.

Dewey and the Deliberative Scientist

Dewey argued that the most effective pedagogy was one that would harness the natural curiosity and impulsive energy of children to more systematic practices of engagement with other people and the world around them. In his Laboratory School, for example, Dewey and the teachers continually provided challenges for the children to overcome, carefully monitoring their activity and imperceptibly guiding them along paths of discovery that led them slowly toward established forms of disciplinary knowledge.

Dewey's commitment to this "scientific" approach to problems was grounded in a key distinction between conscious, deliberative thought, and habitual, routine action. Dewey argued that most of the time people are immersed in the flow of their activity. Drivers can drive cars, for example, only if they do not consciously "think" about every motion they make. Our learned habits, then, often represent capacities for quite complex and flexible engagements with our environment. But, as I noted in Chapter 2, they are limited in their ability to respond to novelty. When we encounter a problem that our habits cannot handle, we must emerge from the "flow" and consciously explore possible solutions. When the gearstick will not move, for example, a driver may need to shift from habitual activity to conscious deliberation about what to do. Only after the problem is solved (by changing the car or one's own driving or both) can one reimmerse oneself in the activity of driving. Conscious deliberative thought in Dewey's model, then, is a critical tool that allows agents to alter either the world or one's own habits or both so that engaged activity becomes possible again, allowing actors to shift back into the "flow."

To a large extent, a Deweyan approach to education is designed to provide students with strategies for fostering and managing this necessary transition between routine and conscious action. While Dewey certainly understood that both forms of engagement lie on a continuum and that people can operate on many different points on this continuum, Dewey's educational theory generally framed these two states as relatively distinct. And Dewey was most interested in fostering *more* conscious, critical deliberation, which he believed had been lacking throughout human history.

Goodman and Neill: The Performance Artist

Central to Goodman's vision of an authentically mature human being, especially in *Gestalt Therapy*, was a capacity for healthy "spontaneity." This involved a complex dance of engagement with one's environment, neither "directive nor self-directive, nor . . . being carried along essentially disengaged." Such a person is "discovering-and-inventing as one goes along, engaged and accepting." The actor is "not merely . . . artisan nor its artifact but growing in" the situation. The primary experience is one of absorbed engagement because a "timid need to be deliberate" leaves one unable to fully engage with novelty.[41]

Goodman understood that human beings cannot realistically avoid dealing with the routine aspects of their day-to-day lives with "automatic" habits that conserve "time and energy," and like Dewey, he knew that "conscious attention" is often necessary to "deal with what is novel and nonroutine." He understood the importance of employing the kind of focused, conscious deliberateness Dewey championed when dealing with especially difficult challenges. But both Goodman and Neill were most interested in teaching people how to dwell in the space *between* the extremes of habitual and conscious engagement. They celebrated possibilities for creative engagement amid immersion "in" activity. Thus, they promoted a developmental path different from Dewey's. They rejected what they saw as the excessively distant, objective "deliberateness" of adults in modern society, which they believed fed the technocratic aspects of their culture. Instead, they championed the "spontaneity, earnestness and playfulness, and direct expression of feeling" characteristic of young children.[42]

What they were arguing for was a form of *controlled* "spontaneity" that remained organically engaged with the shifting complexities of one's context—a kind of moment-to-moment experimental attitude. "Act spontaneously," Goodman advised in an effort to explain this necessary balance, only if you can "be, by study and reflection, the kind of person who can trust himself to act spontaneously."[43] While this approach involved some level of self-consciousness and reflection, it was a limited "healthy deliberateness" that "is aware of certain interests, perceptions, and motions [only] in order to concentrate with a simpler unity elsewhere." Thus, "the primary experience" was a kind of flexible, "absorbed engagement."[44]

If Dewey's fully enlightened individual often shifted into the always somewhat distancing deliberative, experimental stance of the everyday "scientist," then, *Gestalt Therapy*'s ideal person might be described as a "performance artist." The performance artist remains holistically engaged with the fullness of his or her environment, spontaneously but sensitively intervening in and adapting to the moment-realities he or she encounters.

Given this different conception of a healthy person, the free school emphasis on unconstrained self-development makes a lot more sense, reflecting personalists' convictions that such a graceful dance with the world cannot be taught but only allowed to emerge in spaces sheltered from the damaging influence of a technocratic society. In other words, their almost fanatic defense of the "natural" was not just a simplistic ideology of purity. Instead, among more sophisticated thinkers at least, it represented one part of a larger integrated set of strategies meant to nurture a very particular kind of self in their children, a self that would help them resist the neurotic and alienating tendencies of modernity. It was a calculated response to the forms of oppression they believed dominated their society.

Becoming Social: Collaborative Democracy versus "Healthy" Engagement

These different visions of the goals of education and human development also led to different visions of democratic engagement.

Becoming Deweyan Citizens

If schools wished to participate in the development of a truly democratic society, Dewey argued, they must initiate students into three broad kinds of social practices: the shared practices of different communities and disciplines, the practice of intelligent deliberation, and capacities for democratic collaboration. In the case of the former, because of the disciplinary specialization and spatial isolation of modern society, schools must provide opportunities for students to learn the discourses and rules of the dominant communities of their time. If they are to nurture a more democratic society, however, schools cannot only teach such preestablished social practices. Students must learn how to intelligently critique and evolve these practices. And schools must actively teach students how to collaborate on such efforts.

Again, Dewey understood that democratic skills for intelligent collaboration developed in individuals as the result of *both* innate capacities *and* learned practices. While he argued that collaborative democracy would not emerge naturally, he also acknowledged the inherently social, cooperative tendencies of human beings. In his major work on education, for example, he emphasized that from their earliest moments what is most important to children are the "expectations, demands, approvals, and condemnations of others," since, fundamentally, "a being connected with other beings cannot perform his own activities without taking the activities of others into account. For they are the indispensable conditions of the realization of his tendencies." Whenever a child "moves," then, "he stirs" others' activities "and reciprocally."[45]

In fact, the capacity to learn in the first place is fundamentally dependent upon these instinctual tendencies for social interaction. One first learns to speak and act as others do only because these practices "are first employed in a joint activity." Dewey argued, "Similar ideas or meanings spring up because both persons are engaged as partners in an action where what each does depends upon and influences what the other does . . . Understanding one another means that objects, including sounds, have the same value for both with respect to carrying on a common pursuit." Thus, children are born "marvelously endowed with power to enlist the cooperative attention of others" and "gifted with an equipment of the first order for social intercourse." Only through these innate capacities is it possible to become part of the human social world in the first place. But just as the curiosity and experimental tendencies of adults are progressively deadened by the strictures of society, "few grown-up persons retain all of the flexible and sensitive ability to vibrate sympathetically with the attitudes and doings of those about them."[46]

As in the case of the experimental attitude, these observations bring Dewey's vision much closer to that of the personalists than is commonly acknowledged. At points like these, Dewey seemed to at least raise the possibility that if developing children were sheltered from the deadening effects that accompany initiation into contemporary society, their innate tendencies might actually lead children to naturally develop into cooperative, "sympathetic" adults. The point is not that they can be left totally to their own devices, but that a particular kind of socially nurturing but minimally directive context might allow such an emergence. Whether this would happen (and what exactly such "shelter" would look like), of course, was a pragmatic, experimental question Dewey never delved into.[47]

Despite these hints, Dewey ultimately rejected the possibility that a democratic attitude might simply emerge in the absence of active efforts to teach it. He attacked, for example, the tendency of followers of Rousseau to look to "nature" for useful ideas of freedom, terming this a "political dogma" that reflected a desire to rebel "against existing social institution, customs and ideals" without carefully thinking through what an adequate alternative would look like. "Merely to leave everything to nature," he complained, simply "negate[s] the very idea of education," trusting to "the accidents of circumstance."[48] Furthermore, he frequently pointed out, the emergence of democracy in whatever form was, in fact, a rarity. Even when it had existed, as in some ancient Greek city-states, it rarely reflected the kind of widespread equality and collective participation in the development of a better society for everyone that Dewey strove for.

For these reasons, among others, he declared that "the notion of a spontaneous normal development of these [social] activities is pure mythology."[49] "Community life," he argued, "does not organize itself in an enduring way purely spontaneously. It requires thought and planning ahead." While Dewey understood that democracy, like scientific thinking, is grounded in the natural endowment of human beings, then, collaborative democracy remained, for him, an essentially artificial social form, a form that must be actively chosen and consciously developed. This was especially true in schools, where he believed the apparent refusal by "false" progressives to acknowledge their responsibility to mold the social development of youngsters threatened to derail possibilities for the development of a truly democratic society. The tendency for children to jostle against each other, a tendency he asserted was endemic in the "new" schools, was the result of each student only having interest in pursuing his or her own particular projects. This antisocial culture represented "a failure in education, a failure to learn one of the most important lessons of life, that of mutual accommodation and adaptation."[50]

Goodman and Neill: Natural Freedom

In contrast with Dewey, the free schoolers put their faith in a particular conception of "natural" development. Goodman, Neil, and other free schoolers envisioned a healthy community as a place where people honestly presented their real selves to each other, relatively unencumbered by banal rules and preestablished ways of being. In healthy communities, they argued, people are free to express, at every moment, their real feelings in response to the uniqueness of every moment of their existence.

Goodman's gestalt theory provides a sophisticated way to describe what this might look like in more detail than was often provided in most free school literature. In fact, a central aspect of the practice of gestalt therapy involved helping clients learn to authentically express themselves to each other. While gestalt therapists did work with individual clients, the central practice involved groups and not individuals. Through group therapy, gestalt therapists sought to restore "face-to-face community and reclaim traditional" communal values.[51] Therapists prodded participants to release their "animal . . . individual self," and to reveal "the social pressures raging war within personality." Participants were encouraged to spontaneously express themselves to others, not like battering rams but instead as part of sensitive responses to the shifting nuances of the situations and relations in which they found themselves entangled. Instead of trying compulsively, deliberately, to control everything around them, gestalt therapists argued, people who learned to release their creative potentials would find that "if things are let be, they would spontaneously regulate themselves" through the natural interaction between self and others. Slowly, a new form of community could emerge involving "a dynamic unity of need and social convention, in which men discover themselves and one another and invent themselves and one another."[52]

On a more concrete and practical level, this was also Neill's approach at Summerhill. Like Goodman, Neill believed that children's developing impulses should simply be allowed to play themselves out. This belief, however, did not eliminate the need to respond to children's complex needs, something Neill often did. In fact, he was famous for his sensitivity to the particular concerns and issues of specific children. But he emphasized that one must always respond in ways that do not affect children's freedom to act. In general, he argued, "the ultimate cure (for war) is the releasing of the beast in the heart of mankind . . . from the beginning."[53] In fact, Neill's favorite response to a child who was acting out in some way was to reward the acting out until the child simply got tired of it. He would pay a child to wet the bed or to lose things. Sometimes he would even join with children in their "bad" actions. If a child was breaking windows, Neill might come along and happily break a few windows with him or her. One time there was an outbreak of swearing at Summerhill, so Neill held a swearing workshop. In these cases, Neill stated, children quickly become tired of the problematic activity. Many, for example, came to the first swearing meeting, but only one child attended the second.

Neill's assumptions about the natural state of human beings mirrored those of many other free schoolers—not surprisingly, since he was the key example they followed. Neill stated that an effective school requires "complete belief in the child as a good, not an evil, being." Bad or difficult children, he argued, are simply unhappy children. He believed that "a child is born a sincere creature." "Self-regulation," he asserted, "implies a belief in the goodness of human nature; a belief that there is not, and never was, original sin." While children might be naturally egoistic when younger, "altruism comes later—comes naturally—if the child is not taught to be unselfish." Ironically, "by suppressing the child's selfishness, the mother is fixing that selfishness forever." These "natural" tendencies, it is important to note, however, could only work themselves out correctly in the quite artificial environment created at Summerhill. Thus his only significant interventions with children were generally therapeutic ones. For many years he held what he called "private lessons" with struggling children, helping them understand the psychic source of their difficulties. Later, he decided that most of this was unnecessary. Most of the time, in the environment of freedom at Summerhill, he found that children would naturally develop in a healthy, "free" manner.[54]

The key organizing principle of Summerhill was what one might term a "natural consequences" approach.[55] If a child annoyed other people, then he or she had to deal with the effects of his or her actions. Neill noted, for example, one time when some children messed up some potatoes he had planted. If he had made a fuss about how it was a bad thing, *in general*, to mess up someone else's project, he would have been being authoritarian. He would have been trying to fit children into some abstract system of morals and social rules. On the other hand, it was OK to get upset about "my spuds" as long as he "did not make it a matter of good and evil." In other words, it was wrong to mess with his spuds "because they were my spuds and they should have been left alone." While there should be no general rules for behavior, in a world with many individuals with different desires and needs we will continually encounter situations where we will necessarily need to moderate our actions on a situational basis. "No harm is done," he stated succinctly, "by insisting on your individual rights, unless you introduce the moral judgment of right and wrong. It is the use of words like naughty or bad or dirty that does harm." Through their continual interactions in a community where they would repeatedly come up against the needs and desires of others amid an adult-secured environment of positive regard, Neill believed that children would grow up happy and successful on their own terms. In fact, he included an entire chapter in *Summerhill* on the often nontraditional lives of his children after they left the school, focusing on how happy they were and on their ability to achieve at quite high academic levels (as doctors, mathematicians, etc.) if they so wished.[56] Many, not surprisingly, ended up in caring professions.

The aim at Summerhill was to develop happy, whole, creative, "free" individuals. He sought to escape the focus of traditional education where children "have been taught to know, but have not been allowed to feel." He wanted to help children become "creators" who "learn what they want to learn in order to have the tools that their originality and genius demand."

Like Goodman, he wanted to help children gain the "ability to work joyfully and to live positively."[57]

Summerhill was a community designed to care for the needs and interests of unique individuals, and the culture of Summerhill, even its few collective rules and its democratic structure, were designed to foster this end. Famously, the one avenue of governance at Summerhill, aside from the limited authority of the teachers and Neill, was the weekly school meeting where issues were brought up and voted on. Every member of the community, no matter how young, held an equal vote. This was explicitly an effort to teach children both civic skills and civic responsibility to each other and to the larger community. For example, the meeting was always run by a different "chair," and children quickly learned that "the success of the meeting depends largely on whether the chairman is weak or strong."[58]

Like Dewey, then, Neill accepted and modeled the need for structures to help politics work effectively. However, the focus of the meetings were quite different than those of Deweyan democracy. Dewey wanted children to be continually engaged in, and working to manage, common projects that solved shared problems for the entire community. In contrast, the focus of Summerhill weekly meetings was on the conflict between individuals and the group. While the meetings did develop rules for the collective, these usually seem to have focused on how to draw reasonable boundaries between individual freedom and individual rights. Much of the discussion at these meetings revolved around questions about how to respond to the desires and infractions of individuals. The "perennial problem" that Summerhill always struggled with was not the development of community as a collective project, but instead "the problem of the individual vs. the community."[59] When Neill declared that "when the child's individual interests and his social interests clash . . . the individual interests should be allowed to take precedence," he was not, from his perspective, being antidemocratic.[60] For he and most of those who followed his vision believed that it was through the valuing of individual interest that a democratic community would naturally emerge.

Like other free schoolers, Neill focused on the need for individuals to overcome their immersion in the anonymous, bureaucratic structures of modern society. If individuals were able to express themselves honestly to others and responsive to the unique needs and concerns of those they interacted with, he and the free schoolers believed that democratic communities would "naturally" emerge. If children could be released from neuroses and provided with a healthy community where people responded honestly to each other, children were perfectly capable of developing a healthy, flexible sociability responsive to their own unique inclinations without much further guidance. In fact, therapeutic intervention was only necessary, especially for adults, because the depredations of modern technocracy had damaged individuals' natural tendencies for sociability.

In other words, unlike Dewey, who expended enormous energy exploring how a democratic community might be nurtured through sophisticated pedagogical practices, the most influential free school scholars, like Goodman and Neill basically did not worry about democracy as a coherent project. An organic kind of democracy in which individuals (as individuals and in groups) responded to the unique perspectives and desires of others was the natural result of an environment that allowed children to be "free." And, they argued, they had the evidence to prove that this could be successful in Summerhill and in many schools that looked to it in spirit if not in all the specifics. Democracy, for them, was not about collaboration on common projects but responsiveness to the unique perspectives of others within egalitarian communities. The key goal was not contributing to social improvement but the creation of rich communities of authentic interpersonal interaction.

Social Change and the Political

Dewey and the Development of a Democratic Society

From the beginning, Dewey saw collaborative democratic education as a key tool for the creation of a truly democratic society. As I have showed in earlier chapters, however, it remained unclear in Dewey's writings exactly how the democratic practices he valued might actually empower children. In fact, teachers in the Laboratory School reported that children found that the social practices they had learned proved of very limited usefulness in a world where everyone else was not inclined to collaborate in the honest, open manner they had learned to expect. Further, it was not clear exactly how a democratic society built on a Deweyan model of engagement might actually operate. As we have seen in *The Public and Its Problems*, he could not coherently imagine how the forms of relatively intimate collaboration that he preferred might effectively guide communities much larger than the classrooms he had worked in, to say nothing about an entire nation. Despite this, Dewey remained committed to this general vision of a democratic society, hoping that someone might somehow find a solution to the problems of scale that he had been unable to solve himself.

In the end, then, while Dewey's analysis of authentic collaborative democratic practice was enormously sophisticated, his reasoning about its connection to social change and democratic politics remained quite limited. He ended up, in *The Public and Its Problems*, looking to vibrant local communities as a critical base for a new Great Community but was unable to conceptualize how these locals might somehow coalesce into a pragmatic structure for governance and broad social democracy.

Goodman and Neill: Political Change and Democracy

Goodman and most free schoolers, for their part, were just not that interested in more standard or structured forms of politics. However, as with the Young Americans almost a half century before, they still held a sincere if vague commitment to democratic social change. As Goodman and his colleagues acknowledged in *Gestalt Therapy*, for example, "the task" of therapy "would . . . be immensely easier if we enjoyed good social institutions."[61] As the personalist writer Theodore Roszak observed, "the life that gestalt therapy leads Goodman to consider healthy is clearly not livable in . . . [the] existing social order."[62]

Like the Young Americans, Goodman and other personalists during the 1960s sometimes acknowledged that any return to the "local," simply recreating local, face-to-face village communities of the past, was not a solution to the larger societal problem of technocracy. Mostly, however, they did not worry too much about broader social solutions, focusing instead on what seemed immediately possible: recreating vibrant local enclaves where individuals might flourish more authentically. Partly as a result, personalist discussions about how a more democratic society might emerge were usually extremely sketchy. Goodman, for example, mostly tossed off general statements like this: "If people had the opportunity to initiate community actions, they would be political; they would know that finally the way to accomplish something great is to get together with the like-minded and directly do it."[63] Overall Goodman's was "straightforward anarchist thinking in the tradition of Peter Kropotkin" whose vision of "mutual aid" was very compatible with the general focus of free schoolers on the development of local democracies "that emerged organically out of the sensitivity of individuals to others." Goodman's "anarchist sociology did not provide a model of the good society but an *attitude* toward political and economic life, one that was communicated more readily by example and rules of thumb than by logical analysis or the systematic elaboration of plans and protocols."[64] Ultimately, Goodman declared, "he was not a political person and had no utopian vision. All he wanted was 'that the children have bright eyes, the river be clean, food and sex be available, and nobody be pushed around.'"[65]

As Jay Miller noted, this vision fit well with the countercultural thinking of the time. Many counterculture participants believed "that massive change in consciousness could only come from an inner revolution of values and lifestyle that demonstrated its superiority to the old system." "Radicals" like Goodman generally presented their personalist "theory as a recipe for *personal* liberation," unlike "Dewey ... and his closest followers" who "were primarily interested in promoting a more rational" and democratic "*social* order." Many believed that "democracy would 'come naturally,' that is, emerge organically, in intimate, caring communities." Goodman was one of many who saw the free schools as political "in the deepest sense" because they allowed the development of "whole" human beings within carefully constructed "counter-institutions."[66] Despite the continuing importance of the anti-Vietnam movement, the emergence of the free schools movement in the second half of the 1960s was part of a broader shift of many middle-class youth away from overt politics and toward a more individual and communally focused "counterculture."

This focus on individual liberation also infused the more overt political activities of white students in the North in the 1960s. The work of Students for a Democratic Society (SDS), for example, often looked more "personalist" than "collaborative." With few exceptions, SDS's efforts to mobilize poor white communities through the Education Research and Action Project (ERAP), were largely failures.[67] They failed, in part, because the white, middle-class SDS leaders were so focused on responding to the unique perspectives of every participant that they often lost a clear focus on achieving a common goal. Their meetings could last for long hours or even days, sometimes grappling with seemingly minor issues like, at one point, whether to take a day off and go to the beach. An iconic photograph in James Miller's book about SDS, *Democracy Is in the Streets*, shows a late night meeting where everyone seems to have fallen asleep around the core leader. But the leader's bright eyes gaze intently into the camera lens, still ready to facilitate more dialogue.[68]

Middle-Class Privilege: The Jonathan Kozol Critique

The fact that the free schools focused on the plight of the middle class was very much a part of the dialogue about the free schools movement in its later years. The most famous, comprehensive, and I think painfully accurate critique came in Jonathan Kozol's 1972 book *Free Schools*, which eviscerated the free schoolers for their almost complete avoidance of the challenges facing poor people in the United States. In blunt language, he complained that many "Summerhill-type schools . . . pretend to abdicate the very significant and important power which they do possess and do continually exercise upon the lives of children, most significantly, of course, by placing them, to begin with, in this artificial context of contrived utopia within a world in pain—a context within which they can neither hear the cries nor see the faces of those whose oppression, hunger, desolation constitute the direct economic groundwork for their options." He acknowledged that "the beautiful children do not *wish* cold rooms or broken glass, starvation, rats, or fear for anybody; . . . [but neither] will they stake their lives, or put their bodies on the line, or interrupt one hour of the sunlit morning, or sacrifice one moment of the golden afternoon, to take a hand in altering the unjust terms of a society in which these things are possible." One of his deepest concerns about the free school approach, then, was that it was "conspicuously and intentionally antipolitical." He complained, "At best, these schools are obviating pain and etherizing evil; at worst, they constitute a registered escape valve for political rebellion. Least conscionable is when the people who are laboring and living in these schools describe themselves as revolutionaries."[69]

He took proponents to task for assuming that pedagogies designed for the privileged would also serve those on the bottom, echoing later criticisms of whole language and other ill-planned constructivist educational efforts by Lisa Delpit, among others.[70] Unlike children of the middle class, he pointed out, "the poor and black, the beaten and despised cannot survive the technological nightmare of the next ten years if they do not" gain the content and cultural skills they need in their schools.[71] He ridiculed intellectuals like Friedenberg who "ward off black and Spanish-speaking kids from universities and colleges, on grounds that they might undergo the same cooptive dangers and the same risk of domestication that white students do."[72]

In many ways, Kozol's critique mirrored Saul Alinsky's concerns, discussed in the next chapter, about the antipolitical tendencies of the attitudes of "liberal" intellectuals. Kozol's main point was that the pedagogies and social assumptions of the free schools movement wore only the trappings of critique while actually serving the interests of the status quo. They facilitated the continuation of inequality and oppression by masquerading as a revolutionary movement while, at the same time, failing to teach even the most basic tools for effective social action.

Lichterman and the Limits of Personalist Politics

At this point, it seems useful to return to Paul Lichterman's case study of the political limitations of "personalist" politics, discussed in Chapter 1. His book provides a sophisticated analysis of relationships between social class and "personalist" practices.[73] Lichterman studied different local activist organizations in California, including two "Green" environmental activist groups, and found that the Greens embodied much of the "personalist" attitude toward social action that I have been describing. The Greens spent significant time in their meetings listening to each other's personal stories, "rarely creat[ing] short, sketchy self-accounts." In fact, saying little about oneself "produced hesitant glances beckoning the speaker to say more." As one leader said, "It's all about *individual empowerment.*" In fact, like SDS's participants in ERAP a generation before, the Greens were so committed to individual empowerment that they had difficulty acting as a collective at all. Because different individuals tended to pursue different efforts, they could rarely bring their entire group's "power" to bear on a single common project. They seemed mostly unwilling to constrain their own desires and perspectives to enable the achievement of shared goals. Overall, reflecting this book's larger argument, Lichterman found that the Greens' personalist politics arose from the middle-class "lifeways" of participants, even though they believed their way of acting represented a universal model for authentic democracy.[74]

One group, the "Ridge" Greens, actually wanted to develop a stronger common purpose, but their commitments to individually focused, personalist practice seemed to prevent this. Eventually, the group dissolved. Other work has told much the same story about similar "personalist" political efforts. In her book on early antinuclear direct action groups, for example, Barbara Epstein found the same tendency to disintegrate for essentially the same reasons given by Lichterman.[75]

Disagreements within a Shared Theoretical Framework: A Common Continuum of Commitments

What I have attempted to show in this chapter is that Dewey's vision and that of the free schoolers were very similar in a number of perhaps surprising ways. In the end, differences between Dewey and the free schoolers turn out mostly to consist of disagreements about the *goals* of education and not about *how* human learning takes place. Dewey simply focused on producing a different kind of person than did 1960s personalists: an analytical scientist instead of a performance artist and a collaborative partner instead of a caring, fellow community member. As in the 1920s, this disagreement about what kind of person education should foster was linked to different understandings of the nature of oppression in modern society. In both eras, the personalist approach mostly sought to address the challenges of the middle class in a world of all-encompassing bureaucracy. Dewey and his fellow collaborative progressives, in contrast, were most concerned about the "social problem" of class inequality, holding up a middle-class model of social practice as a solution. To a lesser extent during the 1920s and more consistently during the 1960s, the personalists simply were not focused on material and social inequality. Instead, they worried most about how to nurture individuality and creativity. And at different times both the 1920s personalists and free schoolers like Goodman acknowledged these very differences between their aims and Dewey's. In other words, not only did they diverge from Dewey, but they understood this and were, at points, able to present clear justifications for the divergence.

In my discussions of SDS and the Greens I showed that concrete differences in capacities for action emerge out of commitments to personalist rather than Deweyan "democratic" progressive practices. To repeat, personalists focus on nurturing individual expression, often sacrificing capacities for joint action, while those engaged in Deweyan collaboration are more prepared to moderate their individual perspectives and contributions in service to common efforts.

From a broader perspective, however, given the real limitations of Dewey's model, in a practical sense these differences may not matter as much as Dewey thought they did. As I noted in Chapter 4, Dewey never figured out how relatively intimate collaborative contexts might inform governance on a broad scale. And he and other collaboratives were unwilling to address the reality of power and conflict in the modern world. In fact, as I argued earlier, it seems likely that collaborative democratic education actually ended up preparing not political activists but students who could work effectively in the middle-class settings of corporate America. Much the same could be said of the free schools. It is not clear, then, exactly what this difference between personalist and democratic approaches to progressive education implies in a practical sense for democratic politics or empowerment through education more broadly. Neither side seems particularly effective as a base for political action and social change.

In the end, despite Dewey's own later, somewhat quixotic engagements in broader politics, both Dewey and Goodman focused on enhancing local, face-to-face communities, convinced, as Theodore Roszak noted, that "only a social order built to human scale permits the free play and variety out of which the unpredictable beauties of men emerge." For both, "only a society possessing the elasticity of decentralized communities . . . can absorb the inevitable fallibilities of men." As Dewey had in *The Public and Its Problems*, Roszak, along with other personalists of the 1960s, concluded that, if we are to achieve authentic democracy, society must "scale down selectively our leviathan industrialism so that it can serve as handmaiden to the ethos of village or neighborhood."[76] Goodman "produced from these elements a theory of what can best be described as anarchosyndicalism that he came to identify," I think rightly in many ways, "with Dewey and progressivism." In the end, for all practical purposes, Goodman's vision and that of the free schoolers actually looked much like Dewey's with its "combination of utopianism and pragmatism."[77]

Conclusion: We Are Mostly Personalists Now

One of the reasons I have spent so many pages describing the emergence of personalism and the theoretical assumptions that underlie it is because I believe that this is actually the approach to community that currently dominates in "progressive" classrooms. While a Deweyan model of *individual* learning, best expressed by a general "constructivist" approach to knowledge, continues to dominate in progressive teacher education contexts,[78] his model of collaborative democracy is much less evident.

Instead of collaboration on common projects, what comes to the fore in most progressive settings today is what Nel Noddings has best described as "caring," the most important current version of personalist theory. As Noddings has carefully described, caring involves helping individuals learn to be fully responsive to the unique perspectives, needs, and desires of those around them. A caring classroom is one in which people listen to others and seek to support others' personal and academic development.[79] A dialogue in an English classroom about a book, for example, rarely involves a Deweyan effort to contribute to a common project or interpretation. Most of the time, in progressive classrooms, it is a space in which every individual is encouraged to express and defend his or her opinion and to challenge and encourage the perspectives of others. In this way, classrooms can generate quite warm and embracing communities that support the primary purpose of education in America today—the enhancement of individual capacity. This empowers individuals, especially individuals who arrive equipped to participate successfully in middle-class discourse practices. Caring interactions can create a kind of "democracy," but do not lead to sophisticated capacities for developing common projects, even on a local level. And, as Dewey feared, these kind of classrooms are often wrongly represented as examples of Deweyan democracy.

Without a clear explanation of what personalism is and how it diverges from Deweyan "democratic" progressivism, we can and do call ourselves Deweyans, blissfully unaware that Dewey himself would have rejected most of our communal efforts in his name. Only by seeing clearly what we are doing and where what we are doing came from can we begin to consider whether this shift is problematic.

At the same time, as earlier chapters indicated, we need to understand the limits of Dewey's collaborative vision. To what extent, we need to ask, are the distinctions between more personalist or more collaborative classrooms differences that *make* a difference to students and the society beyond the school? It may turn out that, for many purposes, it does not make that much difference at all. That, at least, is the argument of this book.

Part IV

Democratic Solidarity

6

Community Organizing

A Working-Class Approach to Democratic Empowerment

Previous chapters traced the historical and theoretical underpinnings of collaborative and personalist progressivism, the most influential models of democratic practice in academia. This chapter turns away from these middle-class visions to examine an alternative approach I call "democratic solidarity," drawn largely from working-class experiences and cultural practices: *community organizing* in the tradition of Saul Alinsky.

Alinsky was neither the first nor the last influential "organizer" in America. Perhaps most important, union organizers had been fighting for working-class power at least a century before he was born.[1] More broadly, in a wide range of oppressed communities thousands of mostly unsung leaders have long fought for equality in a range of ways. Nonetheless, Alinsky is critically important, not only because he was the first to codify key aspects of the model in books like *Reveille for Radicals*, but also because he was the first to create an enduring organization, the Industrial Areas Foundation, designed to train community organizers and spread organizing practices more widely. Furthermore, for good or ill, Alinsky-based organizing informs nearly all significant community organizing groups in America today. In any event, it is this tradition that I am most equipped to discuss, having worked for more than a decade with local organizing groups.

Only very recently in education have a small number of scholars begun to explore potential relationships between community organizing and education. But without a clear understanding of how the neo-Alinsky vision differs from more familiar progressive perspectives on democratic empowerment, there is a danger that the dominance of progressivism may lead education scholars to mistakenly reinterpret Alinsky's distinct perspective into more familiar progressive terms.

Outside of education, the literature on organizing is somewhat more substantive. However, most of it is descriptive in character. Very little critically examines the theoretical and philosophical underpinnings of organizing. This chapter begins to respond to these lacks in education and beyond, extending on writings by Donald and Deitrich Reitzes and Heidi Swarts among a few others,[2] but my central goal is to explain community organizing, not critique it. For more critical analyses, see my ongoing series, "Core Dilemmas of Community Organizing," which appears on the blog Open Left.[3]

My Own Perspective

Because a good deal of my understanding of community organizing comes from my own experience, it seems helpful to say something about that experience. When I arrived in Milwaukee, Wisconsin, for my first academic position ten years ago, I had spent most of my time studying fairly abstract theories of human agency and empowerment. Increasingly dissatisfied, I looked for ways to engage more concretely with social inequality. It took me about a year to find Milwaukee Inner-city Congregations Allied for Hope (MICAH), a congregation-based community organizing group. After attending an intense week of leadership training and participating in a few campaigns, I was hooked. For the past decade, I have been a member of MICAH's education committee, working on issues ranging from class-size reduction to school bussing policies to health care in urban schools. And over the last two years I have coordinated a new "community organizing coalition" that draws nearly all substantive organizing groups in our city together for the first time in recent history. These efforts were supported by the unique focus of my department at the University of Wisconsin–Milwaukee, which focuses in part on community engagement and change. I have taught our required course in community organizing over most of my time in the department.

Saul Alinsky and the Birth of Community Organizing in America

> To hell with charity. The only thing you get is what you're strong enough to get—so you had better organize.
>
> —Saul Alinsky, *Reveille for Radicals*

Alinsky was a fascinating figure. He grew up in a mostly middle-class family in Chicago in a segregated Jewish neighborhood where he fought Polish youth gangs.⁴ He eventually became a graduate student at the University of Chicago, working with the leading figures of the Chicago School of American Sociology in the 1930s. As a result, despite his rough demeanor and attacks on "theory," Alinsky came to organizing equipped with the best theoretical and practical research training available from the sociology of his time. As a part of his graduate work, Alinsky conducted ethnographic studies of youth gangs and other aspects of the underside of working-class life, hobnobbing with gangsters and delinquents.

He became increasingly interested in how one might help impoverished communities generate the power necessary to resist oppression, and his earliest organizing effort produced the Back of the Yards Neighborhood Council in the slums around Chicago's enormous slaughterhouses. Alinsky's success in bringing together the many ethnic and religious groups formerly kept apart by hatred and suspicion was an enormous accomplishment and brought him widespread fame. This effort also brought him together with labor organizer Herb March and president John L. Lewis of the Congress of Industrial Organizations (CIO) labor union, both of whom influenced his vision of community organizing.

In 1946, Alinsky published *Reveille for Radicals*, the first book in America to codify the key tenets of community organizing. Around the same time he founded the Industrial Areas Foundation, which supported his efforts to develop community organizing projects around the nation and later provided a base for training organizers and leaders.

Alinsky, Social Class, and Community Organizing

In some ways Alinsky played what, after Fred Rose, I call the role of a "bridge builder."⁵ He had one foot in both middle-class and working-class worlds and could critique each side from the point of view of the other. For example, although he learned a great deal from his mentors at the University of Chicago, he reported with characteristic bluntness (and his usual calculated exaggeration) that he "was astounded by all the horse manure they were handing out about poverty and slums . . . glossing over the misery and the despair. I mean, Christ, I'd lived in a slum. I could see through all their complacent academic jargon to the realities."⁶

His organizing perspective was fundamentally grounded in working-class views of the world, developed during years of research and personal engagement in low-income areas, and explicitly rejected approaches preferred by middle-class "liberal" professionals. He had little time for privileged people with "the time to engage in leisurely democratic discussions" and "to quibble about the semantics of a limited resolution." These people, he complained, did not understand that "a war is not an intellectual debate." Liberals had the luxury of adhering to abstract positions and utopian visions of progressivism because "fights for decent housing, economic security, health programs . . . are to the liberals simply intellectual affinities . . . [I]t is not *their* children who are sick; it is not *they* who are working with the specter of unemployment hanging over their heads; they are not fighting their *own* fight."[7]

Alinsky also felt "contempt for so-called objective decisions made without passion and anger." He believed that "objectivity, like the claim that one is nonpartisan or reasonable, is usually a defensive posture used by those who fear involvement in . . . passions, partisanships, conflicts and changes . . . An 'objective' decision is generally lifeless. It is academic, and the word 'academic' is a synonym for irrelevant."[8] To be relevant, he argued, allies with working-class and poor people needed to be partisan—on the side of those who suffered. This did not mean being irrational, however. Instead, Alinsky recommended what he called "cold anger" that could drive careful and strategic efforts to effectively contest injustice.

In contrast with the tendency of privileged middle-class intellectuals and service workers to avoid conflict and seek consensus through rational negotiation, Alinsky argued in very nonprogressive terms that "a People's Organization is dedicated to an eternal war." He stressed, "A People's Organization is a conflict group, [and] this must be openly and fully recognized. Its sole reason for coming into being is to wage war against all evils which cause suffering and unhappiness . . . [I]n a world of hard reality [a People's Organization] . . . lives in the midst of smashing forces, clashing struggles, sweeping cross-currents, ripping passions, conflict, confusion, seeming chaos, the hot and the cold, the squalor and the drama."[9] "Change," he famously argued, necessarily "means movement. Movement means friction. Only in the frictionless vacuum of a nonexistent abstract world can movement or change occur without that abrasive friction of conflict."[10] Of course, more sophisticated collaborative progressive thinkers like Dewey and Addams understood the importance of conflict. But they did not believe that this "friction" necessarily took on characteristics of "war," even Alinsky's version of this, which was always nonviolent.

Alinsky's stress on the "warlike" nature of social struggle was not simply an abstract assessment of social processes. It was also part of a strategy designed to activate particular aspects of human psychology and working-class culture. He sought to "to rub raw the resentments of the people of the community; fan the latent hostilities of many of the people to the point of overt expression."[11] Polarization between a good "us" and a bad "them," he argued, was the most effective way to mobilize masses of people.

He did not mean to insult the intelligence of participants in his organizations, however. In fact, he worked hard to help leaders understand the complexities of specific issues. Unless they understood these details, victory was unlikely when confronted with an opposition that did. But he argued that while "a leader may struggle toward a decision and weigh the merits and demerits of a situation which is 52 percent positive and 48 percent negative, . . . once the decision is reached he must assume that his cause is 100 percent positive and the opposition 100 percent negative."[12] Polarization, he said, was a fundamental political necessity in the arena of conflict and struggle. Of course, this is a classic approach of other forms of working-class struggle, like union organizing.

Later in his life, during the 1960s, he was incredulous when middle-class student activists mouthed personalist arguments about how it was wrong to help poor people adopt "bourgeois, decadent, degenerate, bankrupt, materialistic . . . middle-class values." While Alinsky understood that the challenges of "technocracy" were important, he did not see this as an excuse to ignore the desire of the poor for "a bigger and fatter piece of these decadent, degenerate, bankrupt, materialistic, bourgeois values and what goes with it."[13]

In general, Alinsky saw the middle class as "political schizoids" who "want a safe way, where they can profit by change and yet not risk losing the little they have." Because they have something to lose, they are not willing to take risky confrontational action. As a result, "thermopolitically they are tepid and rooted in inertia."[14]

Practical Ethnography

Alinsky recommended that organizers, who generally came from the middle class, use the extended ethnographic approach to understanding oppressed communities he had learned from his work with the University of Chicago. When Nicholas von Hoffman came to work with him, for example, Alinsky sent von Hoffman off to a community Alinsky was interested in organizing and told him to send him weekly reports about the status of the community. That's it. It was up to von Hoffman to spend day after day going through the community talking with residents, tracking down key leaders and creating relationships with them, and digging up a range of information about the demographics and history of the community. Action only came after months of such work.[15]

The job of an organizer was to immerse oneself into community life to the extent that one was swept "into a close identification" with it, projecting oneself "into its plight."[16] Von Hoffman later described this process in this way:

> It is a very strange thing. You go somewhere, and you know nobody . . . and you've got to organize it into something that it's never been before . . . You don't have much going for you. You don't have prestige, you don't have muscle, you've got no money to give away. All you have are . . . your wits, charm, and whatever you can put together. So you had better form a very accurate picture of what's going on, and you had better not bring in too many a priori maps [because] if you do, you're just not going to get anywhere.[17]

Later on, even when organizing groups began to draw more organizers from local communities, they continued to recommend a similar "self-study" approach.

Solidarity

One of Alinsky's core insights was that power for people without access to enormous bureaucratic or financial resources was almost always the result of solidarity. This was not only an issue for poor and working-class people, although their lack of privilege intensified the need for a collective voice. In fact, he thought that middle-class professionals were fooling themselves when they focused on having their individual voices heard, as if this would lead to significant changes in the world around them.

But then, as Alinsky repeatedly pointed out, middle-class people were pretty comfortable already. It did not really matter that much to them, in concrete ways, whether anyone actually listened or not as long as they had their say—in academic publications, for example. *Their* children were unlikely to suffer much as a result. Near the end of his life, however, Alinsky turned to efforts to organize the middle class, increasingly convinced that those on the bottom needed allies from the middle if they were ever to generate enough power to foster the change they needed and that the middle class would also benefit if they learned to organize.

Alinsky laid out a fairly consistent general rule about the importance of solidarity for power that people with less privilege are more likely to understand, both pragmatically and culturally. Those with urgent and immediate needs are focused more on the importance of change than on whether their individual perspective reverberates through joint action. Yes, they often desire a sense that they are not powerless as individuals. But power is at least partly, if not largely, experienced through being part of something larger than themselves. Few union members walking a picket line, for example, have any illusions about the impact of their own individual actions. But not showing up not only opens someone up to social sanction. It also means the loss of an opportunity to be part of a collective effort, not only concretely in the dialogue at the picket, but also more generally in the sense of being part of something larger than oneself. Walking a picket line represents a different kind of "voice" than participation in the discursive joint projects imagined by Dewey and other collaborative progressives.

To gain and maintain collective power, Alinsky believed, a group must speak with a united voice. In the real world of social struggle, public intragroup disagreement will inevitably be used against you. In private spaces, one can disagree—and working-class people often disagree with great emotion—but there is rarely room for this in public. The kind of "power organization" envisioned by Alinsky necessarily operates something like a collective "person" in the realm of power. And leaders provide the voice of a power organization.

Alinsky and Leadership

> The only way that people can express themselves is through their leaders.
>
> —Saul Alinsky, *Reveille for Radicals*

> Organizers build community by developing leadership. They help leaders enhance their skills, articulate their values, and formulate their commitments, and then they work to develop a relationship of mutual responsibility and accountability between a constituency and its leaders.
>
> —Marshall Ganz, "Online Organizing Course"

Alinsky was a champion of democracy. Sounding much like Dewey, for example, he argued that when people are "organized, they get to know each other's point of view; they reach compromises on many of their differences, they learn that many opinions which they entertained solely as their own are shared by others, and they discover that many problems which they had thought of only as 'their' problems are common to all." In fact, he argued that the actual decisions of an organization were less important than the goal of "getting people interested and participating in a democratic way." For Alinsky, at its core, the central goal of community organizing was to develop "a healthy, active, participating, interested, self-confident people who, through their participation and interest, become informed, educated, and above all develop faith in themselves, their fellow men, and the future." Because "the people themselves are the future," he stated, "the people themselves will solve each problem that will arise."[18] Alinsky was not simply seeking social changes by any means necessary, then. His *primary* goal was the reinvigoration of democratic participation in America, the restoration of the capacities of everyday people to participate in and feel they had some real power over the forces that affected their lives.

But Alinsky's vision of what a vibrant democracy looked like diverged radically from those of the collaborative and personalist progressives. Alinsky's was a *leader*-based model. For example, he argued that "it is obviously impossible to get all of the people to talk with one another. *The only way that you can reach people is through their own representatives or their own leaders*. You talk to people through their leaders, and if you do not know the leaders you are in the same position as a person trying to telephone another party without knowing the telephone number."[19] The kind of "power organizations" Alinsky sought to build were necessarily dependent on what I will call a "leader function." And when he spoke of leaders, he was referring to people who played very specific kinds of roles: what he called "native leaders." A native leader is someone who is actually recognized by a particular collection of people as representing their interests in one respect or another. They differ radically from the "leaders" of social welfare and other professionally run organizations that increasingly dominate impoverished areas today, since these organizations are usually directed by people who do not really have deep relationships with the local community.

Alinsky's conception of native leaders reflected core characteristics of working-class culture. From his perspective, native leaders' capacities for leadership derived from their rootedness in community, in their shared history, and in their experiential understanding of the lives and realities of those who depended on them. They commanded the respect of others because they shared the "aspirations," "hopes," and "desires" of their groups. They had earned the trust of others and their commitment had stood the test of time.

Alinsky argued that the passions and beliefs of the "people" can only be reached through the "Little Joes who are the natural leaders of their people, the biggest blades in the grass roots of American Democracy." Those who look up to the Little Joes have learned to trust them and can use them as indicators of correct and productive action. Alinsky told of efforts to convince people to join an organization and how they commonly asked, "before I want to sign up I want to know if Joe has signed up." If "Joe" had not joined, then they would wave him off, but if Joe had, then they were likely to ask, "what are we waiting for?"[20]

Native leaders are not simply declared to be leaders by outside forces. Nor do they gain authority through their placement within abstract bureaucratic structures. Instead, they must have a real following within the community.

Democracy and the Leader Function

Dependence on leadership does not mean that people in traditional working-class organizations are simply passive followers. Power is often dispersed in many layered and complex ways through committees and multiple collections of differently situated leaders.

In working-class settings, this layering can be reflected in quite formal structures that maintain and respect different levels of status. Unions are a good example of this, if more formal and less fluid than organizing groups. Simple visions of corporate bureaucracy more familiar to middle-class professionals do not adequately capture this complexity or leaders' dependence on their "followers'" willingness to participate in their visions and plans for action. In Chapter 7, I describe what Aldon Morris called the "formal, non-bureaucratic" structure of African American churches in the South during the civil rights movement, a phrase that I also think captures important aspects of the structure of community organizing groups.

Paul Lichterman's study of a working-class African American "anti-toxics" group he called Hillsviewers Against Toxics (HAT) and a middle-class professional "Greens" organization during the late 1980s and early 1990s again provides good examples of the ways these different tendencies can come together in a specific context. In the working-class, African American group, "members participated ... in order to advance HAT as an organization, and Hillsview as its constituency—not to give voice to individual political will as an end in itself." Its organization was formal, with individuals holding clear titles and playing defined roles. The rootedness of HAT's leaders in the community gave them the legitimacy to orchestrate participation by other members at public forums, coaching them on what they were supposed to say. At the same time, however, "HAT board members wanted an organization in which members could in fact become grassroots leaders" themselves.[21]

Like many hierarchically structured groups, HAT was not always able to nurture the kind of broader participation or maintain the openness to new leadership that its constituency desired. In fact, HAT's dependence on its formal leadership structure often limited participation by others, leaving the organization somewhat dependent on its core leader. As Lichterman showed, too much dependence on the "leader function" for the operation of one's organization can create significant problems. As in HAT, key participants may not be allowed, feel empowered, or be willing to step up into key leadership positions. And core leaders may make questionable decisions that their followers may not be willing to challenge.

In his work with organizations from the 1930s into the 1970s, Alinsky understood the dangers of the leader function. He continually worked to link different levels of leadership within and across groups in an effort to increase the mobilizing power and educational reach of organizations. In fact, he could be quite manipulative in moving particular persons to the fore who he thought would be the best leaders. In many of his projects he created new organizations, block clubs among them, in part to grow new leaders and new organized communities that might contribute to the whole. In more contemporary organizing groups, in less underhanded ways, a key task of organizers is still identifying and supporting emerging leaders.

In a world in which one cannot live without leaders, a key strategy for resisting dictatorship or oligarchy is ensuring that you have *many* leaders at multiple levels. And Alinsky was not fully successful in his efforts to develop new leaders, perhaps because he was more trusting in the representativeness of the "native leaders" that he identified than perhaps he should have been. And his leaders were almost invariably men, also limiting their reach and representation. It was not until after his legacy passed to his followers that some of these problems began to be solved.

Leadership in Community Organizing Post-Alinsky

It is in their conception of leadership that organizing groups since Alinsky's death in 1972 have changed his vision in perhaps the most significant ways. I call this the neo-Alinsky model.[22] With the loss of strong ethnic and civic organizations in urban areas, neo-Alinsky groups are frequently based in churches, one of the few contexts where long-term relationships have been maintained in poor (and even in many more privileged) communities. In working-class churches, pastors and key elders continue to play the role of Alinsky's "Little Joes." Decisions and actions are rarely the result of dictatorial action by these leaders, however. They are subject to a range of pressures within their congregations from parishioners. In fact, new pastors who cannot negotiate the established terrain of power in their churches can find themselves quickly churchless.

Because of the deterioration of deeply rooted organizations in American neighborhoods, neo-Alinsky community organizing groups have also come up with new strategies for building (or rebuilding) the relationships necessary for the leader-based structure of a community-organizing group to function. The key technique is what they call the "one-on-one" interview.

In the ideal, prospective leaders conduct large numbers of these interviews with people in their communities. The interviews are characteristically quite personal as leaders probe for the experiences and beliefs that might drive individuals to participate. After such a personal interaction, despite the somewhat artificial circumstances, a relationship is created that the leader can draw upon later. The leader is no longer a stranger, but someone who the interviewee is at least willing to begin to listen to. Organizers often say, "People don't come to a meeting because they see a flyer, people come to a meeting because someone they know invited them." Without leaders who have relationships with a wide range of potential "followers" (hopefully future leaders themselves), few will show up when you need them.

The aim here, in the ideal, at least, is not manipulation. Instead, by discovering the passions of a range of potential participants, a group of leaders can get a sense of what people in their community are interested in, what issues they might be willing to come together around. Community organizing groups like to say the one-on-one process gives interviewees an opportunity to "make their private pain public," to transform their individual and isolated concerns into power for social change.

Sometimes progressives hear about the one-on-one process and misinterpret it as an example of Deweyan democracy. Yes, this process is deeply relational. But it is fundamentally designed to establish effective *leaders* within communities. Once a group of leaders has completed a large number of one-on-ones with people in their community, they can come together and lay out the issues that "their" people indicated they cared about. With their understanding of the "passions" of their people, they then make informed judgments about the kinds of issues their community would be most likely to support. If they choose something their people are not passionate about, they are unlikely to be able to maintain a campaign over the long term. They are likely to face the natural consequence of a lack of interest. These leaders find themselves in the somewhat tense position of many leaders in traditional working-class communities, holding powers of decision always constrained by the desires and beliefs of their followers. The one-on-one process allows leaders to begin to recreate the position of the "Little Joes" of Alinsky's time, rooting themselves in the hopes and desires of their fellows and building trust over time.

While the constraints of established leadership structures are always an issue to one extent or another, the usual problem in today's community organizing groups is not the *restriction* in the number of those who can lead but the relatively small number who are *willing* to take on the burdens of leadership. In contrast with some union positions, for example, leadership in community-organizing groups is a volunteer activity. Only the organizer, who may advise but does not play a formal leadership role, is paid. This makes the leadership structure much more fluid. The few established leaders are often pulled in many different directions with many different responsibilities. As a result, the most important task of community organizers is the development of new leaders.

In part because of their lack of resources and the limited incentives they can offer to leaders, community organizing groups in the neo-Alinsky model embody many aspects of a "formal" but "non-bureaucratic" model. Neo-Alinsky community-organizing groups generally have a visible formal structure, with a central governing board, cross-organization issues committees, and, in congregation-based organizations, "core teams" in individual churches or subgroups of churches. Especially in issues committees and core teams, however, those who are willing to step forward and lead, who can convince others to follow them, can move relatively fluidly between central and more marginal leadership positions. One can become a key leader without ever taking on a formal title. And it is, in fact, in the issues committees and core teams, where specific campaigns are developed and conducted, that much of the real "action" takes place.

There are two major approaches to community organizing today within the neo-Alinsky tradition: "door-knocking" groups that recruit members one-by-one and faith-based community organizing groups (FBCOs) that organize organizations—mostly churches. The national training and support organization Alinsky formed, the IAF, is now an FBCO, as are most of the other major national "umbrella" groups, including People Improving Communities through Organizing, the Gamaliel Foundation, and the Direct Action and Research Training Center. The only national "door-knocking" group is ACORN, but there are many other more local organizations in this tradition.

Leadership and Democratic Theory

In Chapter 3, I discussed tensions within the Deweyan collaborative model of democracy around the question of leaders. I noted that in *The Public and Its Problems*, "despite Dewey's dislike of 'aggregated' desires and 'group minds' . . . there is a sense in which the officials of a state cannot avoid operating as if something like a unitary collective exists. . . . In essence, public officials are accountable to a collective vision that *does not*, strictly, *exist*." It was because of Dewey's discomfort with such aggregation that he restricted officials of "states" to topics on which the "public" had already largely agreed on. Officials could rule only when their decisions involved issues on which there was little controversy. In Dewey's vision of collaborative progressivism, then, a strong leader function was explicitly ruled out. (Dewey would have been more comfortable with what is sometimes called "transformational" leadership, where leaders play the role of facilitators of collaborative action instead of holding decision-making power in a hierarchy.[23])

This discomfort with leadership is a core characteristic of progressive social action more generally. It was a key issue, for example, during the 1960s in the South in the Student Nonviolent Coordinating Committee (SNCC; see Chapter 7) and in the North in Students for a Democratic Society (SDS).[24] It bedeviled the middle-class feminist movements that emerged soon after, in which "organizational structurelessness bred a peculiarly destructive psychodynamics of leadership trashing." It was embodied within the middle-class antinuclear activism of the 1970s, and Lichterman found it in the Greens groups that he examined in the 1990s.[25] This skepticism about strong leaders has remained strong in more recent writings by intellectuals in the collaborative progressive tradition like Benjamin Barber, who argued in 1990 that "on its face, leadership is opposed to participatory self-government," noting that "one might wish to say that in the ideal participatory system leadership vanishes totally."[26] Like Dewey, Barber only wanted leaders who did not actually "lead" in any substantive fashion.

When groups like these become able to engage in effective action, however, it seems generally the result of the emergence of some level of toleration for "leadership." [27] As Miroff noted, while "skepticism toward leadership safeguards groups like these against the dependence of hero-worship ... it also leaves them ill equipped to understand the pervasiveness of leadership in American political life. More important, it impedes any effort to develop forms of leadership that might foster a more democratic politics in America."[28] More generally, Robert Dahl, among others, has argued that leadership is a central challenge for democracy: "To portray a democratic order without leaders is a conspicuous distortion of all historical experience; but to put them into the picture is even more troublesome ... [The] superior influence of leaders violates the strict criteria for political equality."[29] As a result, Ruciso notes, "the theory of democracy does not treat leaders kindly." In fact, "in many respects, democracy came about as a remedy to the problem of leadership, at least as defined by a long list of political philosophers. Fear of leadership is a basic justification for democratic forms of government."[30] "One of the significant consequences" of the desire to avoid grappling with the problem of democratic leadership, however, has been a "remarkable lack of serious studies of democratic leadership as it is actually practiced in modern societies."[31]

This is not the place to examine the tensions of democratic leadership in detail, something that has been discussed by others.[32] For our purposes it is sufficient to note that from the perspectives of Alinsky and later neo-Alinsky organizers, the idea that democratic power can be created without established leadership and hierarchy is a fantasy. Even Dewey, as I described in Chapter 3, understood the overwhelming evidence behind this assertion. A key difference between progressives and proponents of democratic solidarity, then, is the acceptance by the latter of the necessity of a strong leader function. Neo-Alinsky organizers believe that in a world filled with innumerable unique individuals solidarity is only possible when "the people" can achieve, at particular moments, for particular purposes, a unitary voice.

Roderick Bush (drawing from William Sales) noted that, at their best, leaders playing the leader function, like Malcom X, have been "able to give back to people in a highly refined and clarified form ideas and insights that were rooted in their *own* experiences."[33] Only leaders rooted in this way in a shared community experience, who are engaged in continual dialogue with "the people," can even begin to legitimately claim to represent anyone. And in some sense, what neo-Alinsky organizers have tried to do with their "one-on-one" process is to formalize a process of weaving leaders into a web of community relationships.

Note that this vision of leadership is radically different from that of the "administrative progressives" discussed in Chapter 1. Leadership by administrators was not democratic leadership, and it was not meant to be. Far from being "rooted" in the experience of those they direct, administrators were (and are) explicitly separated from them. In fact, Alinsky often complained about professional administrators of local "community" organizations who had no real connection to the community.

There is always some level of accident in the kind of leader one gets, and thus in how the "caesura" in the "voice" of the people gets filled. For example, Martin Luther King Jr. ended up as leader of the bus boycott in Montgomery in part simply because he was relatively new to town and thus had few established loyalties. (This pattern was repeated across the South throughout the civil rights movement.) But the existence of a pool of effective leaders is not an accident. It is the result of sociocultural settings and processes (often formal leadership development efforts) that have nurtured them. Such settings are part of what Alinsky hoped to create in his "people's organizations" and in the IAF.

Few people have the capacity to effectively play all or even most aspects of the leader function for an entire group. In fact, von Hoffman, one of Alinsky's early lieutenants, argued this can actually be a good thing. "When you do find the all-purpose leader," he noted, "you would do well to beware of him. More often than not his domination leads to organizational despotism." Instead, von Hoffman and Alinsky believed that a grassroots organizing group with a formal, nonbureaucratic structure "demands a variety of leadership talents."[34] A variety of leaders bring diverse connections into the community and multiple ways of understanding the mission and goals of the organization.

This diversity can be part of what preserves democracy and maintains openness to a broad range of new leaders. In the end, however, democracy in such settings depends on internal processes for holding leaders accountable and for supporting honest dialogue across disagreements even as a unitary face is presented in public.

Education

Alinsky was deeply interested in educating people about the realities of power and the organizing process. He believed that "the very purpose and character of a People's Organization is educational" and that "the major task in popular education that confronts every People's Organization is the creation of a set of circumstances through which an educational process can function."[35]

In part, he understood this through the lens of his focus on "native leaders." The "Little Joes," he argued, represent "not only the most promising channels for education, but in certain respects the only channels. As the Little Joes get to know one another as human beings, prejudices are broken down and human attitudes are generated in this new relationship. These changes are reflected among their followers, so that the understanding or education begins to affect the attitudes of thousands of people."[36] Note, however, that while his focus here was on the "Little Joes," the impact he sought was more broadly among their followers.

He believed he had seen real results of his efforts. He reported proudly, for example, about a time when an organization distributed ballots about the kinds of issues people wanted the group to work on and received only a few back, with mostly perfunctory responses. A year later, after the organization had been active, the ballots were distributed again and "a much larger percentage ... was returned." Alinsky noted, "This time the response was as different as day is from night. Instead of brief dispassionate, conventional writing, both sides of the ballot were covered with writing and in some cases the individual had attached two or three additional written sheets. The written observations were made with emotion and force, and demanded that immediate action be taken. The issues described were deep and fundamental."[37]

As with many of the stories Alinsky told in his books, there is likely some exaggeration here. Still, his focus on this issue indicated his deep interest in altering not simply leaders' perspectives and understandings but those of the broadest number of community members as possible. Furthermore, he worried about the fact that his techniques were only able to achieve at most 5 to 7 percent participation in a community (even these numbers are likely exaggerations) and complained about the tragedy of "the unbelievable degree of apathy and disinterest on the part of the American people" that his organizations were meant to address.[38]

He argued education should be tightly integrated into the ongoing activity of an organization. While he did create a formal training program for leaders and organizers that his followers have continued and deepened, it was not through formal instruction but through concrete experience of a myriad of different kinds that he thought the vast majority of participants would mostly learn. He believed organizing is most effective when "popular education becomes part of the whole participating process of a People's Organization," as "the stream of activities and programs of organizations provides a never-ending series of specific issues and situations that create a rich field for the learning process."[39]

In this sense, he saw the role of an "organizer" as fundamentally that of an educator. (In fact, he was quite supportive of popular education efforts like Myles Horton's Highlander Center in the South, even though Horton's approach diverged significantly from his, as I discuss in Chapter 7.) The organizer was to stand back so he or she could see the whole without taking sides on internal battles. It was the job of the organizer to facilitate the desires and hopes of the people, not to push his or her own agenda. The core aim was not to reveal the workings of power abstractly or through some kind of "Freirean" dialogue—although these techniques were used when useful. Instead, the organizer educated through *action*, by encouraging people to confront inequality and using these confrontations to nurture understanding and reflection. Rallies, for example, often included speeches from key leaders meant not only to demonstrate power but also to educate participants. And organizers always engaged leaders in critical reflection on actions after they were completed.

Alinsky sought to create organizations that would change participants' ways of thinking as a result of their ongoing participation, by how they were led to *act*. For example, he argued that "the educational slogan has become: 'Get them to move in the right direction first. They'll explain to themselves later why they moved in that direction and that explanation will be better learning for them than anything we can do.'"[40] "The major task in popular education that confronts every People's Organization," he argued, therefore, "is the creation of a set of circumstances through which an educational process can function." A good organizer "knows that in the event of initial failure he must create a new social situation and induce the individual into it;

once the latter has entered the new situation he must of necessity adjust to it."[41] Alinsky's goal was to change the *situations* within which social practices emerge. Like Dewey in the Laboratory School, he sought to use structured social environments to initiate participants into new ways of thinking and acting.

Alinsky went further, however, and argued that in the absence of power, without the capacity to make change, critical thinking not only was difficult to foster but also could be seen by oppressed people as largely irrelevant. "If people feel they don't have the power to change a bad situation," he argued, "then they do not think about it. Why start figuring out how you are going to spend a million dollars if you do not have . . . or are ever going to have a million dollars—unless you want to engage in fantasy?" It is only when "people are organized so that they have the power to make changes, . . . [that] they begin to think and ask questions about how to make the changes." Thus, "it is the creation of the instrument or the circumstances of power that provides the reason and makes knowledge essential."[42] And, in fact, recent studies agree that providing knowledge about social challenges without also helping people generate the power to change these realities often leads to cynicism and immobility.[43]

Respect for What People Bring to the Table

Progressive thinkers and educators like Dewey and Naumburg sought to fundamentally alter the ways children (and the adults they would later become) thought and interacted with the world from the very ground up. Alinsky, in contrast, sought only to teach what he thought was absolutely necessary for generating collective power. In part this was simply pragmatic. His participants were not under the control of "teachers" for many hours for years at a stretch. Further, he was working with fully formed adults, not small children. But this approach also reflected his vision of what "democracy" should truly look like.

Organizers post-Alinsky have developed a related conception of the difference between "public" and "private" engagement. In contrast with Deweyan progressives, organizers intentionally focus on giving people skills for acting in and making sense of the "public" realm, leaving the vast realm of people's "private" understandings and practices mostly alone. Regardless of who you are in your private world, they argue, when you emerge in public you need to play a particular kind of role that can be learned in much less time than can Deweyan practices of collaborative democracy. Organizers are much more accepting of "split" selves. And this approach limits the kinds of changes that need to happen in "who" you are. The point is not that people do not change in broader ways at all but that these changes are not a core focus of the organizing philosophy.

From Alinsky's perspective, efforts like those of the progressives to fundamentally change the social practices of particular groups reflected a lack of "respect." "After all," he pointed out, "if the organizer believes in democracy," then local traditions and cultures must be given at least the benefit of the doubt.[44] More pragmatically, he noted that "a common cause of failure in organizational campaigns is to be found in a lack of real respect for the people" for their traditions and ways of being as well as for their innate capacities, despite whatever desperate circumstances they may live under.[45] His aim was to provide the tools for people to enter the realm of democratic decision making with their rooted cultural selves largely intact.

His point was not that organizers should have no values at all but that these values should be carefully chosen and respectful of the cultural contexts in which they operated. Only if a people's "programs violate the high values of a free and open society," Alinsky argued, should an organizer stand up against them in some way.[46] Alinsky took this stand himself at times, at one point threatening to organize against the first organization he had created, the Back of the Yards Neighborhood Council, when in later years it began to use its power to keep African Americans out of the neighborhood.

Alinsky argued that the "progressive" educational approaches often used by middle-class social workers and others actually degraded their relationships with poor and working-class groups. Other middle-class professionals might have perceived their nonjudgmental, question-asking approaches to interaction as respectful, since these practices were familiar to them as versions of what they had experienced in seminars and other settings. But when used with those with lower levels of social power, whose culture valued honesty over politeness and whose daily experiences diverged quite significantly from those of middle-class professionals, these techniques could seem controlling. They seemed like a refusal to really engage with nonprofessional community people as equals.

Alinsky told a story, for example, of a time when he met with some Native American leaders. When they made some statements that seemed problematic, he told them that "they were full of shit." Later on, these leaders reported to the middle-class professionals they generally worked with that "when Mr. Alinsky told us we were full of shit, that was the first time a white man has really talked to us as equals—you would never say that to us. You would always say, 'Well, I can see your point of view but I'm a little confused,' and stuff like that. In other words, you treat us as children." Alinsky's point was not that an organizer should go around insulting people. Instead, he was trying to exemplify the tension-filled balance between "honesty and rude disrespect of another's tradition."[47] More broadly, he noted that "a community is not a classroom ... and the people are not students coming to classrooms for education."[48] In other words, the approach of a professional "educator," *especially a progressive educator*, is likely to place community people in the position of "students," only accentuating perceptions of difference and inequality.

Human Nature: Self-Interest

Despite his immersion in the best theoretical and practical sociological knowledge and practice of his time, in part the result of voracious personal reading, Alinsky was not a philosopher or an academic intellectual. Nonetheless, it is possible to discern a core set of beliefs about human nature that informed his vision of community organizing.

Alinsky famously believed that people were motivated by "self-interest." What he actually meant by this was more sophisticated than is commonly acknowledged, however. One way to think of this is to distinguish, as current organizing groups often do, between "selfishness," "self-interest," and "selflessness." On the "selfish" side, you have people who join organizations because participation will serve their own individual purposes in very specific ways. They will gain money, resources, or power for *themselves*. On the "selfless" side, you have people who participate out of a sense of duty and lack any real core motivation of their own. What contemporary organizers argue is that neither the "selfish" nor the "selfless" make good long-term leaders in organizing groups. Selfish people do not have the best interest of the collective at heart. Selfless people do not know what they want, so they are unlikely to contribute real energy or leadership and probably will not maintain their participation over the long term. What you want, instead, are people with a "self-interest," what some organizations refer to as "passion," that provides a core motivation for long-term commitment despite fallow periods, strategic challenges, failures, the long timescale of many organizing campaigns, and the inevitable interpersonal issues that arise.

Examples might include someone with a brother in prison for drug possession who joins a committee on drug law reform or a mother who has lost a child to a drunk driver who joins Mothers Against Drunk Driving. Because individuals are all unique, these "self-interests" can run a wide gamut. Core leaders sometimes are there simply because they are desperately hungry for an avenue to contribute to social change—they do not really care what the specific issue is. Some may simply want to "matter" in the world through the kinds of roles that organizing groups allow.

In his first book, Alinsky told a paradigmatic story about how an authentic understanding of self-interest develops. He told how he got David, a store owner, to join his organization by pointing out how his participation might help him advertise his business. Alinsky then walked across the street and got David's competitor, Roger, to join because he wanted to make sure that "David would not take away any part of his business." At the beginning, then, "their sole interest lay in getting as much advertising, good will, and—finally—as much business as possible. They were present to make a commercial investment." But as they participated in the group's children's committee, they

> were sent into some of the West Side tenements of the neighborhood. There Roger and David personally met the children . . . They met them face-to-face and by their first names. They saw them as living persons framed in the squalor and misery of what the children called "home." They saw the tenderness, the shyness, and the inner dignity which are in all people. They saw the children of the neighborhood for the first time in their lives. They saw them not as small gray shadows passing by the store front. They saw them not as statistical digits, not as impersonals subjects of discussion, but as real human beings. They got to know them and eventually a warm human relationship developed.

As a result, "both David and Roger came out of this experience with the anger of one who suddenly discovers that there are a lot of things in life that are wrong." Alinsky argued that "if they had been originally asked to join on grounds of pure idealism, they would unquestionably have rejected the invitation. Similarly if the approach had been made on the basis of cooperative work they would have denounced it as radical." The result of their education through participation, however, was that David and Roger gained "self-interest": no longer selfish, but certainly not selfless. They became leaders with productive core motivations to participate over the long term.[49]

From the beginning, therefore, Alinsky's understanding of "self-interest" was quite sophisticated. As he wrote just before this story about David and Roger, he believed that "most people are eagerly groping for some medium, some way in which they can bridge the gap between their morals and their practices." He saw his model of community organizing, at its core, as a way to provide people "with an opportunity for a healthy, consistent reconciliation of morals and behavior." His appeals to selfishness and more base aspects of "self-interest" were tools for drawing people in to organizations where they could discover their real "self-interests" and where they could discover that action was not simply "selfless" but could accomplish something real for all.⁵⁰ In the most general sense, then, organizing seeks to make it possible for people to live their values.

Means and Ends

As discussed in Chapter 3, a core reason Dewey clung to his vision of broad collaborative democratic governance despite the extensive evidence indicating that it would not work was his understanding of the tension between means and ends in social action. Yes, from where he sat—when writing *The Public and Its Problems*, for example—he could not see how one might pragmatically achieve the "Great Community." But it was always *possible* that future new discoveries might change this, making his dream possible. It was because of this possibility that he could not countenance means for producing social change that would teach actors habits that would lead them away from collaborative approaches to democratic engagement. He was willing to wait out immediate change in hopes that the future might bring solutions he could not then imagine, and he was willing to risk the lives of those who lived in the world as it was and the lives of those that would come after on this gamble.

Again, Alinsky was no philosopher. In many ways, his writings and public statements held philosophers in contempt—those "liberal" thinkers who are so caught up in complexities that they never act. But he understood that concerns about "means and ends" were a key problem preventing middle-class readers, especially, from grasping the logic behind his message. So in his last book, *Rule for Radicals*, he gave a significant number of pages over to a discussion of this "philosophical" question of means and ends.

Like Dewey, as an ethnographer and sociologist, Alinsky understood the power of culture. He repeatedly emphasized that people were inevitably immersed in a social milieu that initiated them from birth into particular sets of social practices. Like Dewey, he knew that radical cultural change was difficult to foster. And he understood that "revolutions" without changes in a group's social milieu and social practices would end up simply replicating the system that had been overthrown, albeit with new people at the top. "To assume that a political revolution can survive without the supporting base of a popular reformation," he argued, "is to ask for the impossible in politics."⁵¹

He nonetheless did not encourage his organizers to attempt fundamental restructuring in the social practices of those they worked with. Instead, by immersing themselves within a particular cultural milieu, organizers were to discern the strategic opportunities already available in particular cultural settings and historical contexts.

The point was not that he did not attempt to nurture any cultural change at all. For real social change to happen he knew that "local groups and agencies must break down their own accumulations of prejudices and feelings, and undergo a period of disorganization in order to make way for the new values and the new philosophies and new purposes."[52] But, again, Alinsky's cultural aims were much more modest than those of the progressives. Instead of trying to turn working-class people into middle-class professionals, he sought to initiate them into a limited set of pragmatic strategies for the development of collective power, always adapted and appropriated to one extent or another to fit the realities of particular contexts and groups.

Alinsky was not seeking to create a utopian democratic society. Unlike Dewey, he accepted what he saw as the limitations of the realities of the world around him. The struggles of the poor and the marginalized against the "haves" were, he believed, likely to be with us *forever*. This was a key fact that he thought "liberals" like Dewey did not understand. In a world where the nonviolent war against injustice would never end, to fail to equip those who suffered with the tools that would allow them to fight was the most desperate folly. What liberals did not understand was that "in our war against the social menaces of mankind there can be no compromise. It is life or death. Failing to understand this, many well-meaning liberals look askance and with horror at the nakedness with which a People's Organization will attack or counterattack in its battles." He believed it was only because of their relative privilege that

> liberals will settle for a 'moral' victory . . . These liberals cannot and never will be able to understand the feelings of the rank-and-file people fighting in their own People's Organization any more than one who has never gone through combat action can fully grasp what combat means . . .
>
> It is very well for bystanders to relax in luxurious security and wax critical of the tactics and weapons used by a People's Organization whose people are fighting for their own children, their own homes, their own jobs, and their own lives. It is very well under those circumstances for liberals who have the time to engage in leisurely democratic discussions to quibble about the semantics of a limited resolution, to look with horror on the split-second decisions, rough-and-ready, up-and-down and sideways swinging and cudgeling of a People's Organization.[53]

The few examples of places where some level of true social equality seemed to have been achieved, for him, would have been the exceptions that proved the rule. They were very rare, and they generally occurred in places (e.g., Scandinavia) where social difference was at a minimum. He would have warned that such spaces were always tenuous. Unless people maintained a tradition of social struggle, achievements could be taken away at any time.

From my perspective, the most important rule about "means and ends" in *Rules for Radicals* was number ten: "you do what you can with what you have."[54] When you are suffering, you need to pragmatically look for ways to relieve this suffering. Yes, you must be concerned with what will come afterward. But to watch your children suffer because you cannot find the adequate moral "means" is simply not a realistic option in the real world. The true horrors are those many situations when there are, in fact, no apparent "means" for relieving the suffering of one's group. Many groups have faced this situation, from generations of black slaves in America to the Jews in the Holocaust. The need for "correct" means must always be balanced against the need for social change. When we look to the privileged to find the correct balance on this scale of suffering and moral uplift, Alinsky tells us, we are inevitably going to find the relief of suffering given less weight. This is precisely what we saw in the earlier chapters. Because of their relative privilege, the progressives of different stripes had their thumbs pressed down on the "moral" side of the scale, on the side of reasonable collaboration, on the side of aesthetic self-realization and organic communities, their eyes turned away—without recognizing this—from the plight of those who their preferences inevitably disempowered.

Ultimately, in cases of true oppression one does not often have the luxury of finding the kinds of "means" that fully embody the practices of the "ends" one seeks. Does this mean that one waits to act until such means are available? Alinsky would say no. This, he would argue, is the inevitable tragedy of the limitations of the condition of the oppressed. The point is not that he embraced revolutionary actions that end up simply recreating the status quo with a new set of oppressors. But in his writings for a mostly middle-class audience, he stressed the importance of not trying to decide "for" the oppressed when it is the right time to act or what the "right" way to act should be.

Class Tensions in Contemporary Community-Organizing Groups

Despite his complaints about middle-class liberals, near the end of his life Alinsky became increasingly convinced that it was only in alliance with the middle class that poor people would be able to develop sufficient power to achieve significant social change. He did not live long enough, however, to engage in any sophistication with the challenges that arise when middle- and working-class people are brought together around the table.

Chapter 1 discusses the dynamics behind this tension between working- and middle-class cultures and discourse patterns in detail, giving examples of the ways working- and middle-class approaches to organizing differed. The chapters that follow, including this one, trace the emergence of these different approaches to social action and ways of framing social problems in more detail. When they come together, working-class groups more comfortable with hierarchy and less inclined to demand space for individual expression meet middle-class groups that have learned from childhood to make their voices heard. This leads to conflict between established groups, as I discuss in Chapter 1.

These problems are only intensified when *individuals* from these groups try to work together within the *same* group. Unless there is extremely effective facilitation, training, or structure, when middle-class professionals try to work with working-class people the professionals tend to take over and working-class participation fades. For example, a congregation-based organizing group I know of made the decision about a decade ago to seek out churches outside of the inner city. One can trace the beginnings of a decline in black, working-class participation in the group from that decision. Similarly, in his work with multiracial and multiclass groups, Eric H. F. Law found again and again that "the white members of the group would disclose their insights and thoughts verbally and freely while the people of color would just sit and listen." It seems helpful here to give more space than usual to his thoughts. Law explained that those from more privileged groups in our society

> participate as they always do, and talk when they have something to say. If they disagree with someone, they disagree with them verbally and openly. Pretty soon, they realize that some others are not speaking. So, with all good intentions, they try to include them by giving subtle hints because it is not considered polite to put people on the spot. . . .
>
> The more they try [to be inclusive], the more the people of color close up. As a result, they [privileged participants] make decisions without the input and concordance of the people of color members, even though they appear to have consented to it. Then, the people of color get blamed for not participating. Occasionally, some white members feel guilty about dominating the group once more.
>
> Those from less privileged groups, operating on a leader-based model of organization take part in the group by expecting an authoritative leader to tell them what will happen and what to do. Instead, they hear many people talking without being invited to speak first. The assumption then is that these people must have a great deal of power and authority; so they let them

talk and do not challenge them. Then, the white members of the group start hinting that they should be talking also but without a direct invitation.... When the meeting ends, they leave and refuse to come back again.

These practices "are implicitly learned and are very difficult to change." They happen on the internal cultural level—on the instinctual level where the parties involved are not even conscious of why they feel the way they do. Since each person thinks only in her own thought pattern, she cannot even understand why the others do not perceive things the way she does.

To some extent, training people "out" of these discursive patterns is to try to train them not to be working- or middle-class in quite fundamental ways.

One of the key points of this book is that the different class-based practices I have described each have their own strengths and limitations. In an abstract sense, it seems logical that groups would benefit from integrating the different progressive and working-class models together. For example, one might want to encourage Deweyan collaboration in small group efforts to develop common issues to work on and then shift to forms of democratic solidarity for more public action. While working-class groups are not necessarily nondemocratic or anti-individual in any broad sense, they may lack aspects of the dialogic practices developed by collaborative and personalist progressives.

However, it turns out that this kind of practical integration is difficult to do. In part for the reasons noted by Law, it is hard to separate these different practices from the privilege of different class positions. In our society, these different ways of speaking and being are deeply intertwined with questions of power and assumptions about the place of individual perspectives in group action. When middle-class people come into working-class spaces, they bring with them core assumptions about what authentic "democratic participation" looks like and often fail to acknowledge the coherence of working-class visions of democratic solidarity. And middle-class people can seem arrogant and impractical in their desire to look at everything from multiple perspectives.

Bridging these two ways of being is no trivial matter. If, as Fred Rose's studies indicated in Chapter 1, it is difficult to get different working-class and middle-class associations working together on common projects, it is even more difficult to get working-class and middle-class individuals to work together in the same organization on any substantial level of equality. As a result, Law noted, middle-class people who have "tried to be 'in solidarity with' the poor" have generally "found their efforts fruitless and frustrating." Even when they are "equal in terms of economic and political power, their attitude and behavior based on their difference in perceptions of power still separate them."[55]

The first step (especially in educational settings), is for middle-class professionals—those with the most individual power and self-assurance—to acknowledge the benefits and strengths of working-class ways of life. Until this happens, we are unlikely to see middle-class dominated settings like schools taking advantage of the potential for the development of collective power resident in forms of democratic solidarity like that developed by Alinsky and his followers. Nor are people from working-class backgrounds likely to become welcoming of middle-class practices of collaboration and expression until those bringing them understand the real limitations of these forms of "democratic" engagement. Until representatives of the middle class can critique their own ways of being and find aspects of working-class culture to cherish, we will remain stuck at this impasse. Neither side is likely to learn much from the other. And this is a great loss in a world where the enormous problems we face require as much civic capacity as we can muster in order to respond to them.

Conclusion: Democratic Solidarity

This model of neo-Alinsky community organizing provides a concrete way of conceptualizing what I call "democratic solidarity." Community organizing groups do in fact exist around the nation operating under this model of self-governance. And they have proved extremely effective in many arenas. They show that a more working-class vision of democratic solidarity can be made a pragmatic reality.

As I argue in my ongoing series on "Core Dilemmas in Community Organizing" on the blog Open Left,[56] community organizing as a model, even in its evolved neo-Alinsky form, is far from perfect. Nonetheless, currently it represents one of the most effective sets of pragmatic practices for generating power in impoverished and marginalized communities, something progressive visions of democracy have failed to provide. And the failure of many academics and educators to engage honestly and deeply with the possibilities entailed in this essentially working-class model is an enormous loss, not only for the field, but also for youth in a wide range of communities. Our deep commitment to progressive visions of democracy has resulted in a failure to provide concrete tools for generating power to those who most need them. In schools, especially, but in many more spaces dominated by professionals as well, we have failed to teach people how they might actually generate the collective power necessary for them to have a real voice in our "democracy."

Part V

Case Study

7

Social Class and Social Action in the Civil Rights Movement

An Introduction to This Case Study Chapter

While grounded in history, up to this point this book has focused on a mostly theoretical exploration of relationships between social class and social action in America. This chapter provides a "case study," examining how these social-class tensions played out in a particular context. It looks across three different major organizations in the civil rights movement during the 1950s and 1960s in the South, showing how each drew differently from the practices of democracy discussed in earlier chapters. (Two additional case studies did not ultimately fit in this volume; those who are interested can access them at EducationAction.org.[1])

The aim of this case study is not simply to show that the more abstract theoretical arguments of earlier chapters are reflected in actual contexts. I also explore some of the rich ways the abstractions I discussed earlier can become transformed in real contexts of social action beyond schools.

The Civil Rights Movement in the South

The civil rights movement remains the most important and well-known example of collective social action in America. Those interested in promoting social change and resistance to oppression almost invariably draw examples and metaphors from this moment in history. But general ideas about nonviolence and Martin Luther King Jr. have had a tendency to accrete together in a hazy, reassuring cloud of positive energy. With the exception of a few historians like Daniel Perlstein and Charles Payne, education scholarship, especially, lacks much sense of exactly how resistance in the South played out or of the myriad strategies for social action that were employed by different groups at different times. As Michael Dyson has argued more generally, in many ways the story of the civil rights movement has been co-opted by a cultural establishment intent on transforming radicalism into a comforting tale.[2] The key problem for education, in particular, is that our tendency to produce uncritical and analytically shallow celebrations of the civil rights movement has denied us access to important lessons this moment can teach us about the tensions between different approaches to collective resistance.

From the perspective of this book, perhaps most problematic has been the tendency of many education scholars to understand the civil rights movement as an example of Deweyan discursive democracy.[3] Maxine Greene's work has been influential in this respect. In works like *The Dialectic of Freedom*, Greene implied, at least, that civil rights efforts constituted complex, instantiations of the kind of discursive collaborative spaces imagined by Dewey (or similarly, if in less detailed fashion, by Hannah Arendt). By overfocusing on a Deweyan model, Greene and others unintentionally perpetuated a kind of middle-class cultural imperialism, focusing on the particular aspects of the struggles of civil rights pioneers that most resembled their own cherished progressive vision of democratic collective action and obscuring others that did not. Despite many laudatory aspects, I have argued elsewhere that Greene's work is often affected by just the kinds of class bias I have discussed, reflecting the field's general blindness to our social and intellectual positioning.[4] Interpretations like these allow us to believe what we already want to believe about social action, indicating that progressive practices of democracy can provide a sufficient basis, by themselves, for empowering pedagogy.

The reality on the ground in the South during the 1950s and 1960s was much more nuanced and complex than non–civil rights scholars generally acknowledge. In truth, the movement encompassed a wide range of organizations and groups with different approaches and understandings of what it was they were trying to accomplish. Even internally, there was often tension about exactly what model of "democratic" social action to follow. It was in fact this fluid diversity—of theoretical frameworks and practical approaches—that differentiated it as a "movement" in contrast with the kinds of centralized, institutionalized neo-Alinskyan "organizing" efforts discussed in the last chapter. "Movements" are generally made up of loosely connected collections of participants, groups, and organizations pursuing a never entirely defined common project. Movement leaders are as likely to be responding to the unpredictable actions of co-leaders and "allies" as they are to the responses of those they oppose.[5]

From the perspective of class practice, three organizations emerge as paradigmatic: Martin Luther King Jr.'s Southern Christian Leadership Conference (SCLC), the Student Nonviolent Coordinating Committee

(SNCC), and Deacons for Defense. While the SCLC was led by middle-class, mostly educated pastors, at its core it was very much animated by the kind of working-class forms of democratic solidarity explored in Chapter 7. SNCC, in contrast, sought to embody a more fully democratic "bottom-up" model of collective action. Ironically, however, SNCC's model was deeply indebted to educated, middle-class practices of discursive democracy, intermixing aspects of personalist and democratic progressivism. The least well known of these groups, Deacons for Defense, coalesced later than the other two out of indigenous, working-class fraternal organizations. Unlike the others, its key leaders *and* its mass of participants came from the working class, broadly conceived, and the group itself was grounded in working-class approaches to social action.[6]

Given constraints of space and time, I present fairly "broad brush" descriptions of these three groups in the discussion that follows. My aim is not to capture the full sophistication of each of their approaches but, instead, to lay out the core commitments and practices that animated each one. I examine the ways their cultures embodied aspects of and tensions between the different models of social action presented in earlier chapters

As part of this effort, I address social characteristics beyond social class—especially gender and race. Given the inescapable fact that nearly all groups focused much more on racial than economic oppression, limiting my focus to social class would inevitably distort the complexities of the movement. Expanding my angle of vision also allows me to challenge the limitations of the tendency, in earlier chapters, to treat class as a distinct "variable." As I acknowledged early on, while it is sometimes useful to treat social class analytically as a separate category, in actual contexts it is always the case that class takes on unique meanings depending on the ways it is intertwined in other aspects of social life, including local culture, racial oppression, and gender. Substantively, as opposed to analytically, there is no pure category of "class" in separation from the innumerable aspects of social life at any particular moment of history. Nonetheless, my analysis continues to emphasize the special influence of class culture on choices between different approaches to social action.

Tensions between Purity and Pragmatism in SNCC

The pioneering work of SNCC in the heart of the Southern Black Belt, perhaps the most racially oppressed region in the United States in the 1960s, has often been held up as one of the best examples of collaborative democratic resistance on an essentially Deweyan model. Here I argue that it is more accurate to see their approach as a unique admixture of personalist and collaborative progressivism, leavened with aspects of local working-class cultural practices.

Formed in 1960 out of a meeting of student groups concerned about racial oppression, SNCC was carefully nurtured until the middle of the decade by Ella Baker, a longtime activist with contacts in local communities across the South. Unlike King, who saw himself and a small group of other middle- and upper-middle-class preachers as leaders of a mass resistance movement, SNCC members developed a very different sense of themselves as catalysts for the emergence of local leaders and locally led movements.

Baker was convinced that "charismatic leaders" like King, who "could rally an anonymous mass of followers to turn out for . . . events," were not really engaged in community organizing. Without fostering local leadership and organization, she felt "mobilizing" efforts like this had little chance of truly rooting themselves into communities for the long haul. Her aversion to charismatic leadership became widespread in SNCC. SNCC members would often refer to King as "the lawd" and resented it when King arrived as a kind of savior in communities they had been hard at work organizing.[7]

Popular Educators

From the neo-Alinsky perspective of the last chapter, it is somewhat inaccurate to describe the SNCC field staff as "organizers." To the extent that they followed Baker's lead, they really played more of the role of progressive popular educators. Baker "viewed a democratic learning process and discourse as the cornerstone of a democratic movement," continually emphasizing that "common people were capable of identifying the problems they faced and learning how to address them." In her vision, the roles of the SNCC field staff were as instigators and question askers; in contrast with Alinsky-based organizers, their job was not to train local people in specific strategies. Field secretaries sought to encourage local people to see themselves as actors with the capacity to make change in their world, facilitating democratic processes to help them develop their own strategies for change.[8]

SNCC members were also informed by their experiences with the most famous popular educator in the South during these years: Myles Horton and his Highlander School. They often visited Highlander and helped to spread the Citizenship Schools developed by Highlander across the South. Like Baker and other SNCC members, despite his folksy style Horton was highly educated. In the 1930s, Horton had returned to the South from New York City to create Highlander. In New York he had worked with a range of major intellectuals of his time, including Reinhold Neibuhr and Dewey. Horton drew from a range of different sources, including his own early efforts at community education and visits to Folk Schools in Scandinavia. And his vision was especially indebted to Dewey, with whom he kept up a long-term correspondence. In Highlander workshops, participants were encouraged to contribute their multiple perspectives to shared problems. By asking strategic questions, facilitators encouraged the development of possible solutions that participants tested against their own experience and often through role-playing. At the end of a workshop, participants were asked what they planned to do when they returned to their communities and were encouraged to come back at a later date to process the results of their actions and plan further engagements.[9]

An Aversion to Class Hierarchy

Baker's support for local people was rooted in an explicitly class-based aversion to hierarchy in the African American community. She "despised elitism" exemplified by the form of organizing embraced by King's SCLC and the National Association for the Advancement of Colored People (NAACP) before it, in which educated, middle-class ministers or staff directed the actions of the common people. Unlike the SCLC, she "rejected the notion that the black middle class had special claims on leadership of the black community." This was a radical idea at the time, since middle-class blacks often viewed their "uncultured" working-class brethren with a measure of disdain. While she "appreciated the skills and resources that educated black leaders brought to the movement . . . she urged SNCC organizers to look first to the bottom of the class hierarchy in the black community, not the top, for their inspiration, insights, and constituency." She "taught the young people in the movement who had achieved some level of formal education that they were no smarter, and certainly no better, than the uneducated farmers and workers in the communities where they were organizing."[10]

SNCC organizers soon learned that this focus on the bottom as opposed to the top of the African American community was not just an issue of democratic ideology, since "in general, anybody who had a specific economic tie-in with the white community could not be counted on when the pressure got hot." Only those without much to lose were likely to push for action (ministers, with their relatively independent economic base, were a key exception among the middle class). According to Baker's Biographer, Barbara Ransby, "the resulting inversion of the conventional class hierarchies within SNCC was most pronounced. Talented and educated young black people were persuaded to forfeit their privileged claim to leadership of the race, a status that would naturally have been afforded them, . . . and instead to defer to the collective wisdom of sharecroppers, maids, and manual laborers, many of whom lacked even a high school education."[11]

Middle-Class Practices of Democracy

Baker's stress on the limitations of class hierarchy conflicted to some extent, however, with the fact that in its early years SNCC's most influential staff members were highly educated. Robert Moses, SNCC's most respected member and a crucial model for other SNCC organizers, for example, came to SNCC from a stint at Harvard graduate school studying analytical philosophy. He and Jane Stembridge, another early SNCC staff member, first met and became friends by "talking about religion and philosophy [and] reminiscing about their common East Coast ties." In their discussions, "they debated the writings of philosophers Paul Tillich and Albert Camus and pondered their relevance to the realities of the agrarian South."[12] While Baker herself had not attended graduate school, she was something of an autodidact and had no trouble holding her own in discussions with staff like Moses and Stembridge. In fact, the "leaders" of SNCC were much more educated than most of middle-class ministers who led the SCLC. Thus, there was an enormous gulf between leading SNCC staff and the local people they engaged. And while they intended to let local people think for themselves, SNCC members seem to have had a very clear sense of the kinds of practices that were most "authentic." In sometimes subtle ways, SNCC seems to have championed social practices of engagement derived in large part from the experience of middle-class intellectuals that were somewhat alien to local people.

Baker's vision of democratic engagement, for example, was essentially one of a network of face-to-face personal relationships. She was not much interested in mass movements. "Democracy," for her, "was about fairness and inclusion, not sheer numbers . . . [and] had to revolve around real participation and deliberation." Combining aspects of personalist and Deweyan progressivism, Baker focused on the empowerment of individuals within community. She was determined "to allow each individual to make a contribution and play a role in his or her emancipation." What needed to be reformed, Baker emphasized, were not abstract organizations but "the human beings involved. Individuals had to rethink and redefine their most intimate personal relations and their identities." She imagined the development "of individuals who were bound together by a concept that benefited the larger number of individuals and provided them an opportunity for them to grow into being responsible for carrying out a program."[13] Overall, like many of the personalist progressives discussed in previous chapters, SNCC "was less interested in executing a well-planned agenda than in enacting in its own operation the society it envisioned." And like Dewey, Baker argued that "there should be no distinct intellectual leadership. Rather thinking and analysis should be incorporated into all aspects of movement work."[14]

While Baker urged her followers to seek out local leaders and to learn from them, her personalist commitments showed in her continual expression of discomfort with significant "leaders" of almost any kind. "Strong people," she famously asserted, "don't need strong leaders." She believed that "when ordinary people elevate their leaders above the crowd, they devalue the power within themselves."[15] While she was all for careful planning, and while she understood pragmatically that different people needed to be responsible for different duties, she nonetheless had an aversion for hierarchy, an aversion she transmitted to SNCC. Moses similarly held tight to convictions honed in his broad philosophical reading. As his biographer, Eric Burner, noted, Moses "embodied ideas and these, not leadership in itself, were his passion." Moses struggled with the tension between pragmatic action and ideological purity, fundamentally believing that "active participation is as much end as means." Like Baker, he focused on efforts to allow individuals to act together in groups without suppressing their unique identities. His work "was akin to the purpose of Camus's outsiders to heighten consciousness. Only by doing so could individuals decide for themselves what choices to make." In the end, Burner concluded that Moses's "work in the South during these early years suggests a self-contained young philosopher hoping that each of the people he reached could be similarly centered." By presenting himself as an example for others, however unintentionally, he sought to encourage others to be more like himself.[16]

Inspired by Baker's vision of "group-centered leadership," SNCC leaders "opposed any hierarchy of authority such as existed in other civil rights organizations"[17] and resisted, at least in the early years, any effort to centralize power. This "determination" by the more educated members of SNCC, including Baker and Moses, "to allow each individual to make a contribution and play a role in his or her emancipation . . . informed the creation of a fluid structure" so that SNCC operated without clear lines of authority or a stable sense of "who" was in charge. And they brought this antiauthoritarian vision with them into the field. This aversion to leadership and hierarchy came along with a discomfort with the development of strong and stable institutions. Very much a Deweyan pragmatist in this way, Baker "saw revolution as a process, as a living experiment in creative vision and collaboration, very little, in her opinion, could be predetermined." This meant that now and in the future "no blueprint could be rigidly adhered to." And institutions, by their very nature, seemed to embody such limiting blueprints. "Inherent in Baker's philosophy," Ransby argued, "was the recognition that no organization should last forever. Each must yield to something new as historical circumstances changed."[18] In many ways, they saw SNCC as an "antiorganization."[19]

While in an abstract sense it is true that organizations need to be flexible, in practical terms, as neo-Alinsky organizers emphasize, the failure to build strong and durable institutions can also end up allowing hard-won achievements to dissipate. This creates an enormous burden on individuals in communities to continually recreate themselves and their ways of interaction to respond to new challenges without any solid structure to support this activity. Without strong institutions, public spaces for collective action threaten continually to dissolve when participants, for whatever reason, periodically pull away during inevitable fallow periods for action. The belief of SNCC's early intellectual leaders that, in Stembridge's words, "as soon as local community begins to emerge," they should "get out of the community, so that the leadership will take hold and people will not continue to turn to you for guidance"[20] had the potential to hamstring groups before they had fully formed. As more contemporary community organizers have learned, and as Alinsky warned at the time, having local leadership is not the same as having an effective institutional structure to sustain a movement.[21]

Mobilizing the Common People

Because she shared their educational capital, Baker was able to model for her charges how to overcome the gulf between SNCC staff and local people without talking down to those with less formal education. After decades of work with working-class people, first in Harlem and then as an NAACP field secretary in the South, Baker developed "a way of appealing to ordinary people by making herself accessible, speaking in a familiar language that people could readily understand, and interacting with them in a way that made them feel they were important to her." Instead of telling SNCC staff how to organize, she modeled organizing by embodying it in her interactions with them. She often guided them with questions, avoiding simple directives. She showed them "the importance of patience and process," the power of listening more than speaking, and convictions that embodied her faith that those they worked with had the intellectual strength within them to find the right path.²²

Baker also "trained" SNCC staff by sending them out into the field to work under the mentorship of a broad group of strong local leaders with whom she had built relationships. These local people demonstrated to these students "first hand the willingness, ability, and determination of oppressed people to resist and overcome their oppression while speaking for themselves." She wanted her students to see that her contacts were "just as capable as a Martin Luther King or a Thurgood Marshall."²³

The experience of Moses, who later became a mythic figure in SNCC, is a good example of this. Baker first sent Moses to work and live with Amzie Moore, a local Delta activist who had participated in struggles for decades. With Moore's help, Moses immersed himself in the local culture and made personal contacts with local people. Although he did not end up organizing in Moore's hometown, he eventually started an organizing campaign elsewhere. Moses's approach became iconic for the rest of SNCC with his quiet speech and deep listening, clad in his dusty overalls. Like Baker, he became famous for his capacity to engage respectfully with local people. His quiet strength was so impressive, in fact, that the poor farmers and working-class blacks he worked with in his first organizing effort began "to call him 'Moses in the Bible.'"²⁴ Along with Baker, he perhaps best embodied the esprit de corps of the organization. Not by talking, but by acting, Moses helped create the SNCC "tradition" of the early years, when "organizing meant getting rooted and building personal relationships that could be converted into political ones."²⁵

Integrating Local "Working-Class" Practices

It is important to emphasize, however, that neither Baker, Moses, nor other SNCC staff were political purists. Despite deeply held views, they were also pragmatists, responsive to the needs of particular situations and the perspectives of local people. Despite her discomfort with strong leadership, for example, Baker strongly supported Fred Shuttlesworth, the key local leader in Birmingham. Even though others saw Shuttlesworth as "an authoritarian preacher rather than an inclusive democrat," Baker felt that his "zeal and charisma" were "talents applied in the service of the collective interests of the disenfranchised rather than self-aggrandizing gestures on the part of a single individual." Baker was willing to support him despite her general discomfort with strong leadership because she thought "he tried to act in accordance with the wishes of the 'masses,' even if he did not always poll them" directly.²⁶ Similarly, despite their general dislike of "the lawd," SNCC nonetheless found itself actually pleading with King in Birmingham when they realized they needed his leadership to keep demonstrations going. Even Moses ended up taking strong leadership roles in the organization when this seemed necessary.[27]

Similarly, it is important to acknowledge the real openness of Baker and SNCC members to the power of local culture. Baker continually stressed that if SNCC was to build strong local organizations, they needed "to understand and decode the culture of everyday life," using this knowledge "to tap the reservoir of resistance that resided there."[28] This message was also impressed upon SNCC staff at Highlander. Thus, in the South, middle-class, Deweyan visions of collaboration between unique individuals and personalist visions of individual actualization within organic "beloved communities" met vibrant local practices of group engagement.

Like King's SCLC, discussed later, SNCC found, for example, that they could adapt "a rich church culture to political purposes. Black spirituals, sermons, and prayers were used to deepen participants' commitment to the struggle."[29] These practices often focused less on individual actualization than on public assertions of solidarity and collective destiny, often drawing ideas of "collective deliverance" from the Old Testament. In many of these meetings, "folks were feeling themselves out, learning how to use words to articulate what they wanted and needed[;] . . . they were taking the first step toward gaining control over their lives, by making demands on themselves."[30] They were making sense of the ideas SNCC activists and others were bringing to them through the lenses of their own local cultural ways of acting and understanding.

Local traditions of solidarity also played out in SNCC's larger meetings. Again as in King's SCLC, given the pragmatic limitations of dialogic democracy within such large groupings, meetings often "served as forums where local residents were *informed* of relevant information and strategies regarding the movement."[31] Further, given the incredible danger faced by movement participants, mass meetings often served not only as places to plan or to discuss actions but also as opportunities "to help . . . participants . . . overcome often paralyzing fear and its displaced expression." Amid violence and harassment, engagement in political action was possible only because it was "generally a collective rather than just an individual action."[32]

Similarly, "freedom songs" sung in meetings, during actions, and elsewhere "were less concerned with conveying information or arguing a position than with expressing resolve and public solidarity . . . [They] were particularly striking ways of making a collective presence known to the outside world—and to the participants themselves."[33] More generally, as Payne noted, "in some respects, mass meetings resembled meetings of Alcoholics Anonymous or Weight Watchers. Groups like these try to change their members by offering a supportive social environment, public recognition for living up to group norms, and public pressure to continue doing so. They create an environment in which you feel that if you stumble, you are letting down not only yourself not only your friends."[34]

Middle-class practices of self-expression through storytelling were also often transformed in the context of Southern culture and racism. For example, "an important element in reconstructing the consciousness of ... [participants in the movement] was simply having them publicly recite their biographies." In contrast with what Lichterman found in his study of the Greens discussed in earlier chapters, however, these were often not simply—or even mainly—occasions for self-actualization. Instead, they provided opportunities for individuals to "turn private and individual grievances into a collective consciousness of systematic oppression" even as "individuals created a public face for themselves, which they then had to try to live up to."[35] Individual contributions affected the collective's sense of their shared effort, as Dewey had envisioned, but at the same time this activity often served to link individuals more tightly into shared norms and a sense of collective solidarity.

Finally, while the myriad individual relationships SNCC activists developed with local participants were designed to draw people into collaborative engagement, as Payne noted, "for many in the South, attachment to the movement meant attachment to individuals."[36] Dialogue in this context, then, often linked participants into clear relations of hierarchy and authority at the same time as they promoted democratic engagement. Whether SNCC staff wanted this or not, their actions in local communities transformed them into local leaders. Someone known as "Moses in the Bible," for example, is a leader whether he wants to be or not. And, as I noted, Moses at times accepted this reality. (At one point he tried to reduce his influence by asking people to call him by his middle name, "Parris," but it is doubtful this had much effect.)

Engagements between SNCC staff and local people amid pragmatic efforts to promote change, then, generated a fluid collection of practices. Individual, collaborative, and collective identities became inseparably intertwined in unique ways given local challenges and intersections between local culture and the middle-class culture of activists. Each side learned from the other. In this context, dialogue and relationships often fostered solidarity and a vision of collective democratic action against oppression as much as it served visions of collaboration and individual development. Nonetheless, it is clear that key "leaders" like Baker and Moses held tightly to their personalist and democratic progressive commitments, continually resisting and downplaying the importance of tendencies toward solidarity and especially hierarchy.

Losing Faith in the "Beloved Community"

Despite the bravery of SNCC's field secretaries, by the middle of the decade many had begun to despair about the possibility of social reform in the South and, at the same time, began to question the model of organizing championed by Baker and Moses. On the one hand, SNCC had successfully helped catalyze powerful resistance movements, especially in Albany, Georgia, and Birmingham, Alabama. On the other hand, SNCC staff had seen few substantive changes in the status of Jim Crow in most areas; few black voters, for example, had actually been registered. And they had begun to see that challenging economic issues, as opposed to the mostly legal civil rights issues they had so far focused on, would be even more challenging.

During this time, it is no accident that SNCC began to fracture along class lines. SNCC field staff, increasingly drawn from less-educated Southern youth, began to express differences with highly educated central office staff. In fact, one paper distributed at a staff retreat during this time argued explicitly that the most "pressing problem" for SNCC "was the differences in background of SNCC's college-educated central office staff and its black field staff in local offices." The paper complained that the "field workers were 'closer in backgrounds' to the local people, 'while those making broad policy decisions [were] removed physically, if not also in terms of backgrounds and experiences from the people out in local communities." Southern black staff were increasingly concluding, in contrast, "that they had been too naïve" and began seriously questioning the general underlying assumptions of SNCC's work that "'poor people were good and could do no wrong,' and that 'such things as leadership, money, power, etc. were by definition wrong and were things that SNCC people should avoid.'" SNCC field workers were becoming frustrated with this "party line." They began "to see themselves as local leaders rather than as outside organizers" and were increasingly "disturbed by the tendency of others" in SNCC "to equate a belief in local leadership with a generalized rejection of all authority and institutions."[37]

A split began to emerge within the organization between those termed "floaters," mostly the relatively elite followers of Moses, and "hardliners," mostly Southern blacks with poor and working-class backgrounds expressing a desire for more structured organization and leadership roles. Cleveland Sellers, "a southern-born black organizer whose political education had taken place largely within SNCC" and a prominent "hardliner," embodied the attitude of his fellows in his lack of "interest in the philosophical issues that absorbed his opponents." He denigrated "floaters" with "epithets" like "'philosophers, existentialists, [and] anarchists,'" complaining about the chaos that could be created by those who "believed that every individual had the right and responsibility to follow the dictates of his conscience, no matter what."[38]

These fractures emerged amid a group of exhausted fighters who had been literally crashing against the intransigence of Southern whites for years with very little outside support. Amid the continual tension of their precarious lives, both sides exaggerated the differences between them. Certainly neither side held to the kind of ideological purity attributed by the other. As I have shown, Baker, Moses, and their colleagues were, for the most part, thoughtful pragmatists, continually struggling to balance their democratic ideals with the realities of oppression in the South. They did at times support more directive forms of leadership than their general stance might have implied. Nonetheless, in the face of criticism Moses continued to defend he and Baker's overall vision of a fluid democracy without commanding leaders or strong institutional structures. Moses still asserted, without much actual evidence, "that, freed of outside domination and allowed time to develop, local movements would gradually recognize their common problems and would conclude that people everywhere should join in creating a society in which everyone participated in political decision making."[39] For their part, the "hardliners" seem generally to have understood the tension between organizing and leading local communities and the dangers inherent in usurping local leadership.

Despite some exaggerations, however, SNCC staff during the mid-1960s rightly saw that the divergences between the "floaters" and the "hardliners" were driven, in large part, by differences in the social class backgrounds of each side, manifested most clearly in the educational gulfs and regional differences between Northern and Southern blacks in the organization. What proved to be a fragile and relatively short-lived integration of the approaches of different forms of middle-class progressive and working-class practices fractured under the pressure of endemic racism, violence, and despair.

Middle-Class Leaders and Working-Class Actors in the SCLC

King's SCLC reflected a different relationship between working- and middle-class people in the South. In contrast with SNCC, whose educated leaders promulgated middle-class professional models of social action at the same time as they sought to avoid formal leadership, groups like the SCLC and the myriad local organizations it supported embraced a much more traditional hierarchy with middle-class leaders and mostly working-class followers. The structures of the SCLC's pastor-led organizations, however, were much more nuanced than the SNCC's complaints about "the lawd" implied. They were not simply hierarchical bureaucracies. Because these resistance organizations emerged out of the participatory structures of black churches, they were quite responsive to the myriad perspectives and the convictions generated by the ongoing dialogues of participants. In fact, significant aspects of these groups' power often rested in the hands of committees made up of the rank and file. This reality of shared leadership and dialogue could be obscured by the collective drama of mass meetings. In fact, mass meetings represented only one of many arenas of collective engagement for the movement, specifically designed to present a unified front to a hostile world and to generate a sense of collective purpose and unity. Leaders of these groups were not simply dictators then. The best leaders, like King, were less interested in enforcing their own visions on followers than on embodying and reflecting back on what they saw as the best aspects of the emergent perspectives of the community in their speeches. In nearly all cases, then, they were "native leaders" of the kind Alinsky was always looking for.

The Position of Minister-Leaders

The pressure of parishioners often drove pastors to take on the dangers of their leadership roles against Jim Crow. Frequently, "the enthusiasm of a black population trapped the ministers in their leadership role."[40] Gayraud Wilmore exaggerated only slightly when he noted that "black preachers, most of them still unknown and unsung, were there only as instruments—sometimes the reluctant instruments—from which the theme of freedom rose like a great crescendo from the depths of the people."[41]

It is important to understand that ministers were not born; they were chosen. They had to prove themselves to their parishioners in order to take up the mantle of leadership. "As Reverend Vivian [of the SCLC] explained, ... the black minister had to ... be nominated to that post. He had to be voted in. He had to be made the pastor. Nobody sent him down. ... That minister had to make it out of nothing, what we call the rough side of the mountain.'"[42] Even King, coming from the family of a distinguished pastor, had to prove himself.

In fact, during the mid-1950s—especially in urban but also in rural areas—middle-class ministers' claims to the leadership of the working class were very much in question. Wilmore noted, for example, that there was "probably no place in the world where the Christian church was under a more sustained and demoralizing attack from those who were once on the inside of than in the black communities of the United States" during the decades prior to the civil rights movement. Not only black intellectuals but also working-class parishioners were increasingly unhappy with the privileges of preachers who seemed to contribute little to their communities. Prior to the emergence of the movement, the "black church was a place where the minister could speak of hope but no action."[43] Many, especially men, voted with their feet.

King and other leaders did not simply take up the reigns of religious authority then. Instead, they, King especially, embodied a new vision of the promise of Christianity for making a pragmatic impact on racial oppression. People followed religious leaders not because they were under the spell of an overpowering religious compulsion, then, but because they became convinced through their leaders' actions and words that they were deserving of their leadership positions.

A "Formal, Non-Bureaucratic" Structure

Organizations were able to emerge in the South with such speed because they were grounded in the existing structure and organizational model of the black church. The church "provided the early movement with the social resources that made it a dynamic force, in particular leadership, institutionalized charisma, finances, an organized following, and an ideological framework through which passive attitudes were transformed into a collective consciousness supportive of collective action." As Aldon Morris has pointed out, the traditional organization of the black church is not one of strict "bureaucracy." Instead, it has a much more fluid, "formal, non-bureaucratic" structure. While "behavior within the church is organized," with different committees and levels of leadership, "much of it is not highly formalized." Instead, in each church "the personality of the minister plays a central role in structuring church activities." Thus, "the relationship between the minister and the congregation is often one of charismatic leader to followers rather than the formalized levels of command found in large corporations."[44]

King (among others) was "keenly aware" of the dangers of strict forms of bureaucratic control. This was, in part, because of his and others' experience with the strictures of the NAACP, which had constrained local actions and helped prevent the emergence of a mass movement before the organization was banned across the South. In fact, King intentionally neglected "internal structures so that SCLC's staff could move quickly as the circumstances changed and the opportunities presented themselves."[45]

Ironically, this nonbureaucratic structure was part of what made it appear from the outside that ministers operated like dictators. Baker, among others, complained bitterly about the imperial pronouncements of the minister-leaders of the SCLC. And it was true that the SCLC's lack of structure often impeded its efforts. SCLC leaders often "behaved like a council of barons... given to long winded speeches, ... to deliberative posturing, and to a work process consisting largely of decrees, delegations, and postponements."[46]

Even in the case of the SCLC, however, decisions usually emerged through extended dialogue in often widely distributed arenas facilitated by a range of different leaders and activists. In fact, "without such an arrangement, it is unlikely that the complicated business of managing mass movements could have been accomplished." This was even more the case in local action organizations, which "had definite visions of labor and clear lines of authority," channeling "power ... through various organizational positions and personnel."[47] The apparent charismatic domination of key movement leaders, then, often obscured the extent to which these groups drew on a range of more or less democratic procedures, especially in the context of ongoing local resistance activities.

Mass Meetings

The visibility of mass meetings was a key component that led outsiders to perceive the movement as simply an outgrowth of the centralized power of King and other key leaders. And, of course, mass meetings did not necessarily reflect the kind of egalitarian, face-to-face democracy of small groups. A church packed with five hundred or more participants is simply not an appropriate place for a free-flowing democratic dialogue in a manner recognizable to progressive democrats.

Nonetheless, in ways different from those preferred by SNCC, even SCLC's leader-focused mass meetings had perhaps surprisingly democratic aspects. The crowds often formed long before leaders arrived, organizing themselves without much external direction, with the crowd singing hymns and giving testimony to each other. And aspects of bottom-up control could continue even after leaders took the podium. In Montgomery, for example, "some of the more outspoken old people were moved to speak from the floor." In fact, "their folk wisdom and their tales of daily life inside the homes of powerful white people ... [became] a special treat at the mass meetings, bringing both entertainment and inspiration."[48]

Mass meetings also "functioned not only to build community support for leaders' decisions, but also to prevent middle-class leaders from making secret agreements and compromises with the white power structure."

Despite the limitations of the forms of acclimation allowed in such public spaces, "plebiscitary democracy" at least "guaranteed that all agreements had to pass muster with the black rank and file: the working class, the poor, and the youth."[49]

In general, however, the meetings served a role within the movement separate from the many dispersed spaces within all movement organizations for dialogue and participation. As SNCC also found, mass meetings were opportunities to display and develop collective solidarity. They were generally places not for the assertion of individuality but, instead, for the dissolution of individuality—not in the sense that people abandoned their capacity for individual judgment but instead as opportunities for joining together in a collective struggle with fellow citizens. "The church culture that permeated the meetings enabled the diverse groups (professors, porters, doctors, maids, laborers, housewives, even drunks) to abandon the claims of rank while reaching out to each other in new hope and faith . . . Under the impact of the Old Negro Spirituals, of hand-clapping, shouting, 'testifying,' and 'amening,' personality shells dissolved and reintegrated themselves around a larger, more inclusive racial self."[50] "Mass meetings" were "events/sites where participants could express" shared "fear as well as resolution, anger as well as understanding." In these and other arenas of public self-assertion, participants "were less concerned with conveying information or arguing a position than with expressing resolve and public solidarity." Freedom songs in these contexts, for example, provided useful shorthand ways of "summing up a general stance, and preparing for action."[51] While individuals often gave "testimony" at these and other meetings, telling personal stories, in contrast with the middle class, was often meant to enact a person's embeddedness within a shared experiential framework, something I noted earlier as a common characteristic of working-class storytelling. In contrast with dialogue in middle-class, professional, social-action contexts, these stories were meant less to generate unique "I"s than to embody and extend a shared story of "us."

Charismatic Leadership

An understanding of the multifaceted role of "charismatic" leadership is critical for making sense of the role of ministers in this aspect of the movement. First of all, it is important to understand that even highly educated pastors like King were deeply rooted in the local culture and social systems of the black community in the South. They understood the rich realities of the lives of their parishioners—they had to if they were to maintain their positions in their churches.

At the same time, while on the whole (with the key exception of King) less educated than SNCC leaders, they were "among the best educated of the black clergy" in the South.[52] Their dual positioning, with a finger on the pulse of the local community and access to middle-class practices and knowledge, allowed them to play the roles of mediating figures between black and white, working-class and middle-class cultures.[53] In fact, it was this role that parishioners often demanded they take on. Since many local community members lacked a sophisticated sense of the workings of the world of white power, it made pragmatic sense that the pastors would take charge of broad strategy for the movement. Despite the real accomplishments of SNCC, the vast majority of people in the South were seeking leaders who could make the mass struggle possible.

As heads of nonbureaucratic organizations, the minister-leaders' power rested on their ability to inspire their followers. Their task was not to micromanage the day-to-day workings of protests, although they were often deeply involved. Instead, they worked to develop "strong face-to-face personal relationships" with a range of leaders and actors to whom most of the work was delegated in an effort to "foster allegiance, trust, and loyalty."[54] Despite his regal presence, King was also continually engaged in dialogue with the common people in the movement.

At the same time, in large public meetings and protest events, it was the role of these leaders to "personify, symbolize, and articulate the goals, aspirations, and strivings of the group they aspire to lead."[55] As I noted in an earlier chapter, what Bush (drawing from Sales) said of Malcom X could be said, to a lesser extent, about King and many other major protest leaders. Like Malcom X, King especially, "was able to give back to people in a highly refined and clarified form ideas and insights that were rooted in their *own* experiences . . . [He] was able to obtain and sum up the sense and wisdom of the people" as he worked to 'clarify and refine these ideas, which were based on Black people's historical experience and common culture.'"[56] More generally, "the black ministers of the 1950s knew black people because they had shared their innermost secrets and turmoils." "The words they used were effective because they symbolized and simplified the complex yearnings of a dominated group," spoken from a position "firmly anchored at the center of the ebb and flow of the group they aspire[d] to lead." Their words "were effective because they symbolized and simplified the complex yearnings of a dominated group."[57]

Even more pragmatically, ordinary people needed strong leaders to act. If they rejected King, for example, to whom could they turn instead? Power requires unity and leadership. At the end of the day, they had little choice but to have some level of faith in their leaders. This is a key catch-22 of collective resistance in working-class communities that neo-Alinsky organizers seek to overcome (never entirely successfully) by continually recruiting new leaders.

In contrast with the image of imperial command of a dominated mass of the common people that emerged in SNCC's critiques of King, the evidence indicates that ordinary, working-class folk were quite capable of making their own decisions when they ran up against what they perceived as the limitations of their leadership. Each participant needed to make decisions at every moment about how they would act with respect to each demand made by the movement. And they were perfectly capable of balking when they disagreed with particular approaches and decisions.

In many ways, then, ministers like King took on aspects of the leader roles described by scholars of post-Alinsky organizing. They worked to stay in contact with the myriad perspectives of their followers and sought to integrate these with their own commitments, giving back to them their overarching hopes and desires in more concrete form.

Transforming Individuals or Transforming Structures

While Baker and Moses were well acquainted with the effects of social structures on the thoughts and attitudes of people and communities in the South, their main focus was on the transformation of individual conscience. Mainly, they approached people as individuals and sought to alter their attitudes toward the world. And they tended to downplay the importance of religion and established local church leaders in their efforts to empower the downtrodden.

While leaders of the SCLC, like King, certainly targeted the individual hearts and consciences of followers, they depended on the institutional structure of the church to structure participation. They sought to shift the focus of the church toward more active resistance to oppression, contesting tendencies to embrace otherworldly views of religion and social quietism. The SCLC's minister-leaders understood in a very visceral sense through their experience of ministering that "people's attitudes are heavily shaped by the institutions with which they are closely affiliated, . . . [providing] the cultural content that molds and shapes individual attitudes." It is for this reason, Morris argued, that "changing attitudes by refocusing the cultural content of institutions [engaged in defining social reality] can be much more effective than changing the attitudes of separate individuals," and as a result, "fairly rapid transformation of . . . attitudes may be accomplished" in this manner.[58] This use of the social environment as a tool for transforming individuals was a key strategy also used in different ways by both Alinsky and Dewey.

Interlude on Nonviolence

The idea of nonviolent struggle and the name of Gandhi were not entirely unknown to African Americans by the time of the Montgomery Bus Boycott in 1954. A. Philip Randolph, the president of the Brotherhood of Sleeping Car Porters union, had made nonviolence a central part of his March on Washington Movement (MOWM) in the 1940s, pressuring the federal government to end discrimination in the war industries. Early in the MOWM, he began referring specifically to Gandhi in his speeches about nonviolent action, and he held a conference in 1943 to develop strategies for using nonviolent struggle to attack segregation. In the years after this, the African American press, especially in the North, kept its eye on India and wrote articles about nonviolence for the general public.[59]

Despite Randolph's urging, however, nonviolence as a strategy did not take hold in the wider African American community. While the idea of loving one's neighbor was prominent in Christianity, the truth was that "blacks, like other Americans, were not naturally inclined toward nonviolent action."[60] In the South, even during the civil rights movement, "the concept of nonviolence as a way of life was challenged" in all classes "by the traditional southern black orientation to protect the homestead from white supremacist night riders."[61] In fact, as James Farmer of the congress of racial equality (CORE) later noted, in the years before the civil rights movement, African American leaders and press "were not really interested [in nonviolence] because this was rather a bizarre technique to them. . . . They simply could not see nonviolence. 'No, no, that is just unrealistic. If they hit you, you've got to do something. Hit them back. It just won't work.'"[62]

Strategies and practices for nonviolent resistance did not emerge out of the common traditions of Southern blacks, then. Instead, in the United States these were developed by a small group of mostly white, highly educated activists in the North whose backgrounds and commitments resembled, in many ways, those of the early leaders of SNCC.

The Development of Nonviolent Action Strategies in the North

At first, nonviolence proponents were not focused on racial inequality. The fellowship of reconciliation (FOR), which quickly became the major supporter of nonviolent action in America, hired its first staff to deal with issues of race in 1924. But a focused effort to contest segregation did not come until more than two decades later in 1945 when FOR spun off CORE, which was specifically oriented toward this issue. Under the leadership of

Farmer and Bayard Rustin, among others, CORE immediately began experimenting with nonviolent methods for resistance. In fact, Rustin came up with the idea for the first freedom rides in 1946 to test new Supreme Court rules against segregation on interstate transportation.

Throughout all these efforts, however, CORE remained a relatively "elitist organization."[63] While its concern about racial equality was authentic and while it did engage in real action, the group focused more on developing strategies for nonviolent action than on actually changing American society. In part this was because its largely white, middle-class, and educated membership had limited experience with racial or economic oppression. Overall, then, "CORE was an intellectually oriented organization, concerned as much with broad philosophical issues as with racial equality." As Farmer later acknowledged, "from its early days" CORE focused on "a nonviolent direct action" as a "technique," and the development of these "means" was "perhaps more important in the minds of many of the persons there than the ends which were being sought." In fact, Morris noted that "because CORE was largely led by middle-class white intellectuals during the 1950s, it entered the South with a paternalistic attitude about how poor blacks should fight for liberation." Before the civil rights movement, "very few blacks had the interest, desire, or background required in CORE."[64]

The Nonviolent Strategy Comes South

King was quite familiar with Gandhi and ideas about nonviolence before he became the leader of the Montgomery bus boycott, and from the beginning he urged movement members toward nonviolence. His knowledge, however, was largely theoretical, gained in part through his doctoral study of personalist philosophy. In fact, his house in the early days of the bus boycott was defended with guns. While Rustin of CORE, arriving soon after the boycott began, was "impressed by" what he saw as "the intuitive Gandhian method at work in the plan," he worried that the local organization was "at once gifted and unsophisticated in nonviolence." He immediately called FOR's central office, arguing that "these people *must* have somebody come in who was qualified to teach nonviolence." As a former communist and known homosexual, Rustin was too controversial a figure to become officially connected to the Montgomery effort. FOR sent down Glenn Smiley, a white minister, to work with King. And Smiley began making short presentations about nonviolence before most of the subsequent mass meetings. In the background, however, Rustin quickly became one of King's key advisors. Rustin and Smiley's contributions, along with those of James Lawson, also of CORE, who was training a "nonviolent reserve army" in Nashville, were so critical because this small group included many of those in the United States with the experience and knowledge to teach nonviolence.[65]

Montgomery is often referred to as the first major nonviolent mass movement during the civil rights era, but it did not involve the kind of direct, confrontational action that later dominated news coverage. In fact, the boycott was in large part a strategy of avoidance—blacks avoided engaging with whites by staying off the busses. While there were occasional moments of conflict between individuals and groups—in which blacks largely kept to their nonviolent strategy—the first real moment of mass engagement (and of potential confrontation) with whites came at the very end of the boycott when blacks finally began to ride the busses again. Thus, Smiley and King had many months to prepare people with their somewhat alien nonviolent practices.

Nonviolence and Gender

Acceptance of nonviolent strategies for action in the South often diverged along gender lines. Women were much more likely to accept the approaches taught by emerging experts in nonviolence. In general, they "were more politically active, attended mass meetings in larger numbers, and attempted to register to vote more often than their male peers."[66] In fact, as Lance Hill (and others) have noted, "throughout the South, most black men boycotted the civil rights movement," and as a result, "the campaigns in Birmingham, New Orleans, Bogalusa, and Jonesboro became movements of women and children."[67] The fact that women were more likely than men to participate in church culture may explain some of this difference. Since they were more immersed in Christian teachings about the need to love even one's oppressors, nonviolent action may have made more sense to them. And, in the case of the SCLC, their support of the church may have made them more likely to willingly follow the guidance of ministers.

In contrast, the strongest aspect of cultural life that influenced black men was the Southern code of honor, intensified by the specific forms of oppression and violence they experienced in the South. During the movement, "nonviolence required black men to passively endure humiliation and physical abuse—a bitter elixir for a group struggling to overcome the southern white stereotype of black men as servile and cowardly." At the same time, nonviolence "obliged black men to stand idly by as their children and wives were savagely beaten, a debasement that most black men would not tolerate." In fact, Hill argued that "according to the southern white code of honor, passivity was cowardice and cowards were inherently inferior" and that "southern whites could not respect . . . a man without honor." According to Hill, then, at some points the failure of black men to publicly defend themselves and their families may actually have damaged the social power of the movement in the South. These cultural dilemmas were magnified by the fact that "the physical . . . risks that black men assumed when they joined a nonviolent protest far outweighed what black women and children suffered," although Hill argued that "for most of the black men, the issue was honor, not safety." For these reasons, "nonviolence discouraged black men from participating in civil rights protests in the South and turned the movement into a campaign of women and children."[68]

Many of the men and women who did not participate publicly in the protests nonetheless worked quietly in the background. As Hill and others have noted, "throughout the civil rights movement, African Americans frequently guarded themselves and their communities against vigilante assaults."[69] The histories of the movement make it clear that the appearance of guns and occasional responsive gunfire actually prevented a great deal of violence against activists and local participants. Not only men embodied this defensive attitude; "behind the chivalrous ideal of self-defense . . . lay the reality that women were almost as likely to arm themselves for self-defense as men."[70]

In fact, the hold of nonviolence over participants in different aspects of the movement was always somewhat tenuous. Leaders frequently struggled to maintain this attitude during moments of crisis. The movement's commitment to nonviolence, for example, was moments away from being broken during a famous incident in Albany when King and a large number of participants were trapped by a mob of whites in a church. In the moments before federal marshals finally arrived to disperse the mob, "some of the men who had prepared for this moment were slipping out of the pews, reaching for knives, sticks, and pistols in their coat pockets. There were heated whispers in the wings as some of them told the preachers that they were not about to let the mob burn or bludgeon their families without a fight, even in church."[71] On a smaller scale, moments similar to these occurred throughout the years of the movement.

This especially male discomfort with nonviolence finally erupted in mass retaliatory violence in Birmingham when, after police turned water cannons on the protestors, "young black men, nonpacifists who had previously lingered on the sidelines, now retaliated with bricks and bottles. . . . For the first time in the history of the civil rights movement, working-class blacks took to the streets in a violent protest against police brutality and Klan terror."[72]

The Limits of Nonviolence

Although the ministers of the SCLC and the student activists of SNCC continually worked and sometimes struggled during the late 1950s and early 1960s with locals to maintain the publicly nonviolent character of the movement, some have argued that many of the movement's successes resulted from a general fear that blacks might turn violent. King himself often used the threat of the future riots that might result if nonviolent protesters were not successful to convince powerful whites. Kennedy used the same kinds of arguments to pass the Civil Rights Act.

In fact, Hill has argued that nonviolence actually proved itself to be of limited effectiveness in the first decade of the movement. He speculated, for example, that it was black retaliatory violence in Birmingham that ultimately forced "the first real concession" from the federal government "in the form of the Civil Rights Act." "Only after the threat of black violence emerged," he noted, "did civil rights legislation move to the forefront of the national agenda." Hill likely goes too far with these critiques, however. The very existence of the movement slowly altered black people's attitude toward themselves, toward whites, and toward their own rights across the South. Even though an earlier nonviolent struggle in Albany ended inconclusively, for example, the very existence of the struggle changed relationships between black and white communities. And despite the lack of many concrete changes in the Mississippi Delta during SNCC's years of work there in the early 1960s, it is clear that the actions of these activists and struggling local people did, in fact, transform the self-understandings of many people in these communities. In any case, whether or not Hill and others are correct about the importance of black self-defense to the successes of the civil rights movement, arguments about the limits of nonviolence as a strategy became more convincing to activists and local people in the South as the movement grinded slowly and painfully on.[73]

More generally, scholars of the movement agree that nearly all movement participants saw nonviolence as a strategic response to specific conditions and not as an end in itself. Baker and Moses, among others in SNCC, frequently acknowledged the potential limits of nonviolence. And while SNCC generally encouraged nonviolence, its aversion to hierarchy and leadership meant that they were unwilling to tell local people how they should think and what they should do. In the SCLC it appears that among the Southern leaders only King held an immovable conviction about nonviolence. Hosea Williams noted, for example, that he had "never believed in nonviolence as a way of life" and that he did not "know anybody else" among the Southern leadership "who did but Martin Luther King."[74] Even for King this commitment came relatively late in life. When he spoke of what he had learned in Montgomery at the end of the boycott in 1955, for example, he emphasized that "the lessons of leadership and unity came first, the militancy of the church next, and the 'discovery' of nonviolence last."[75] And even King understood that there were practical limits to nonviolent strategies. Later on, for example, he challenged those who failed "to perceive that nonviolence can exist only in a context of justice ... Nothing in the theory of nonviolence," King argued, "counsels" a "suicidal course."[76] In any case, because most participants held a strategic attitude toward nonviolence, continual dialogue went on at all levels about its relevance, a conversation that became more heated as the 1960s advanced and as the limits of nonviolence became increasingly clear to many.

A Declining Commitment to Nonviolence

By the middle of the 1960s, critiques of nonviolence grew more prevalent across the South. Within the SNCC, for example, according to executive secretary Forman, the experience of freedom summer in 1964 "confirmed the absolute necessity for armed self-defense."[77] By this time, most SNCC staff in the South carried firearms. The character of the movement in the South began to shift toward a more militant promotion of self-defense. Despite King's continual effort to reinforce people's commitment to nonviolence, especially in the cities, both in the North and in the South, his vision was increasingly replaced by a conviction about the importance of self-defense by the varied leaders of the black power movement.[78] At the same time, in the South the threat of violence from whites was increasing as "old style citizen's councils" gave way to groups like the Klan. Across the South, "terrorist violence replaced [the primarily] economic threats [of the Citizen's Councils] as the principal means of social control over blacks." This rise in "white supremacist violence in response to desegregation made armed self-defense a paramount goal for many local black organizing efforts." In fact, during the march to Selma in 1965, King himself actually allowed an armed group, Deacons for Defense, to provide an escort. This was the first time he had made a significant concession to the importance of self-defense since removing the guns from his house during the early days of the Montgomery boycott.[79]

Deacons for Defense

I have focused up to this point on the most famous organizations that worked in the South during the civil rights movement: SNCC and SCLC.

I showed how SNCC was dominated until the mid 1960s by an elite group of middle-class intellectuals whose visions were always in tension with their promotion of middle-class, professional models of group-centered leadership outside established institutions. The leaders of SCLC, in contrast, made no effort to disguise their intention to lead. They and those they spoke for seemed to agree that their education, and at least implicitly their class membership, made them the correct spokespeople for the common people. While SNCC often critiqued what they saw as ministers' heavy-handed pronouncements, the SCLC ministers were, in fact, deeply rooted in the local culture, institutions, and emerging perspectives of working-class people in the South. In fact, as I have noted, it was often the case that the ministers were *pushed* into leadership roles by their congregations. The SNCC and the SCLC, then, represented two different ways that middle-class "leaders" related to the working class during the civil rights movement. Each, in essence, represented a different compromise between local working-class culture and the knowledge and practices of the middle class.

It took nearly a decade after the emergence of the Montgomery bus boycott in 1954 for a local and *working-class directed* organization to emerge in the South. This organization, Deacons for Defense, has been largely slighted in the histories of the movement, an absence now somewhat rectified by Hill's recent (2004) book. With the formation of Deacons, the longstanding but often hidden commitment of many movement participants to aggressive self-defense emerged from the background into the public foreground of the movement. The Deacons grew out of African Americans' long tradition of self-defense in the South and the often quite-systematic network of shadowy protectors that watched over nonviolent activists.

Unlike SCLC, which emerged out of the church, or SNCC, which rejected institutional structures, the Deacons formed within one of the few secular alternatives to the church among African Americans in the South: "fraternal orders" of black men. Long before the emergence of the Deacons, black fraternal orders had participated actively in the movement, organizing protection and providing spaces for groups like SNCC to meet in hostile local areas. In fact, "nearly all of the male civil rights activists" from the South "belonged to one or more of these orders." Before the civil rights movement, in part because they were "free of the constraints of Christian pacifism," black fraternal organizations had been one of the "primary mechanisms for sustaining black masculine ideals of honor, physical courage, and protection of family and community." They provided a space dominated by working-class men independent of the broader white community. It was in these spaces that a palatable alternative to the often evangelical nonviolence of highly educated Northerners could grow.[80]

Unlike both SNCC and SCLC, both leaders and followers in Deacons for Defense were thoroughly working class in orientation. "A unique phenomenon," Deacons was "the only independent working-class-controlled organization with national aspirations to emerge during the civil rights movement in the Deep South and the only indigenous African American organization in the South to pose a visible challenge to Martin Luther King... Reflecting class tensions within the African American community, the Deacons spearheaded a working-class revolt against the entrenched black middle-class leadership and its nonviolent reform ideology."[81] The group formally established itself in 1964, the year when the tide began most visibly to turn against nonviolence as a viable strategy for contesting racial oppression.

It is important to understand that the Deacon's alternative to nonviolence was not "violence." Hill emphasizes that, "in truth, defense groups like the Deacons used weapons to *avoid* violence." Scholars generally agree, for example, that the (publicly unacknowledged) presence of armed blacks throughout the early years of the movement prevented a great deal of bloodshed. Ultimately, "the Deacons did not see their self-defense activities as mutually exclusive of nonviolent tactics and voter registration," believing that "it was quite possible to follow a peaceful path, while rejecting nonviolence's inflexible passive strictures." They used applications of "force" to "coerce change rather than win consent from one's enemies."[82]

Interestingly, the Deacons' vision of "coercion" was actually similar to King's own emerging understandings. As Dyson pointed out, during these same years King was increasingly embracing more militant forms of nonviolence. When he moved north, for example, he "announced a bolder initiative, calling it, alternatively, 'massive nonviolence,' 'aggressive nonviolence,' and even 'nonviolent sabotage.'"[83] And I have already noted King's decision to allow the Deacons to provide protection for the 1965 march to Selma. Differences between King and the Deacons, then, were fewer at this time than one might initially assume.

The structure of the Deacons reflected that of the fraternal groups that it emerged out of and depended upon. Instead of looking to educated ministers or Northern intellectuals, it drew its leadership from the most respected working-class men in the community. It developed a "formal command structure of elected officers" and held to "strict recruiting standards," rejecting those with problematic reputations or those who were known to be "troublemakers."[84] In contrast with the more formless democracy preferred (at least in the ideal) by SNCC, and the fluid "charismatic, nonbureaucratic" structure of the SCLC, the Deacons "adopted a standard meeting format using parliamentary procedure, with the reading of minutes and committee reports." In general, "all major decisions were made democratically," while day-to-day operational decisions were made by elected officers, following their hierarchy of command. In other words, much like the unions they resembled, they depended upon familiar working-class forms of social organization that linked formal operations of democracy to clear hierarchies of control, allowing collective action without constant discussion.

In the end, Hill has argued that the Deacons' ability to combine a range of coercive tactics with the threat (and sometimes the fact) of retaliatory violence extracted significant concessions from the white establishment. "The Deacons' campaigns," he found, "frequently resulted in substantial and unprecedented victories at the local level, producing real power and self-sustaining organizations."[85] Hill argued, therefore, that it was with the emergence of the Deacons that real structural change started to take place in the South.

Conclusion: Social Class and Social Action in the South

> There is a belief, especially within white America, that black people are somehow monolithic as a social and political group. African-Americans know better. The entire political history of black America has been essentially a series of debates.
>
> —Manning Marable, *Black Liberation in Conservative America*

There was no single approach to social struggle during the civil rights movement. Instead, an ongoing debate took place during these years, both within and between organizations, about what the correct balance of approaches should be. SNCC and the SCLC embodied complex, tension-filled, and yet at the same time sophisticated and locally responsive admixtures of different practices derived from class, gender, region, and the like. From the perspective of class, SNCC and the SCLC represented fundamentally different efforts to balance middle-class privilege and education with working-class traditions. While King was immersed in "personalist" philosophy during graduate school, it was SNCC that most clearly embodied the tensions within progressivism between personalist and collaborative democracy. In contrast with both of these groups, the Deacons represented an almost completely working-class-based approach to collective empowerment. Each collection of approaches had its benefits and drawbacks, and likely each was more or less effective given the challenges existing in particular situations and historical moments.

For the purposes of this volume, it is important to emphasize that all the key participants during these years understood and explicitly referred to the fact that social class was a key factor influencing choices for and against different organizational and action models. In fact, I believe that social class was seen, both to participants at the time and by scholars afterward, as the most important social factor influencing choices between and tensions among different models of social action during these years of the movement.

After the middle of the 1960s, the SCLC declined in importance, and working-class organizations became the core of racial struggle in America. The Deacons for Defense are only one example of the growth of broad working-class dominance in the movement after the middle of the 1960s. The SNCC dissolved, parts of it transitioning into the working-class-led Black Panther Party, which was a key participant in the emergence of the mostly urban, working-class black power movement. In all these cases (although there is neither space nor time to address them), key aspects of the more abstract, class-based tensions between different social-action practices described in earlier chapters made themselves felt, even though class alone can help us understand only part of the story of this period.

Part VI

Conclusion

8

Building Bridges?

> The very social scientists who are so anxious to offer our generation counsels of salvation and are so disappointed that an ignorant and slothful people are so slow to accept their wisdom, betray middle-class prejudices in almost everything they write.
>
> —Reinhold Neibuhr, *Moral Man and Immoral Society*

> As an organizer I start from where the world is, *as it is*, not as I would like it to be. That we accept the world as it is does not in any sense weaken our desire to change it into what we believe it should be—it is necessary to begin where the world is if we are going to change it to what we think it should be.
>
> —Saul Alinsky, *Rules for Radicals*

The core message of Saul Alinsky's model of community organizing is that we must act in the world "as it is." He railed against privileged middle-class people who cling to impossible dreams of public "reason" and "collaboration" that have little relationship to the real workings of power in the lives of those who suffer in America. He had little use for middle-class progressives and their efforts to create a democratic society focused on the unique identities of individuals.

Even Dewey could not concretely explain how collaborative progressive practices could form the foundation for governance in the modern world. The personalists did not even try. This book argues, then, that Alinsky's complaints that progressives' tendencies to avoid really dealing with challenges of power and real oppression were clearly on the mark.

It is important, however, to acknowledge the real achievements of collaborative and personalist progressives like John Dewey and Margaret Naumburg. Especially in their educational work, progressives showed that their apparently "utopian" theories could, in fact, be used to create concrete democratic communities. While these community practices were limited in their applications, they nonetheless constituted rich spaces for individual and collective development.

The problem with middle-class progressive visions of community is not that they were useless or unproductive but instead that progressives have generally failed to subject them to sufficient critique. Because both collaborative and personalist models represent elaborated versions of the day-to-day cultural contexts in which middle-class professionals (especially academics) work and live, it still seems intuitively obvious to intellectuals and educators that "authentic" democracy today is equivalent to progressive practice of one form or another.

This tendency to naturalize progressive democracy and equate it with democracy writ broadly has fed into the general tendency of middle-class professionals to look down on poor and working-class people. Across progressive writings, one finds disappointment, as Neibuhr noted, that "an ignorant and slothful people are so slow to accept their wisdom. From this perspective, key aspects of working-class practices like "community organizing" seem undemocratic and primitive. Cooperation rules the day. An exhortation to "fight" brings only genteel disdain.

A central goal of this book is to help progressives move beyond their (often unrecognized) depth of disdain for "others" who do not truly understand how to "be democratic." I seek to help progressives understand the internal limitations of their own most cherished visions of democratic community in the hopes that this will open us up to the capacities of people positioned very differently in our society. With my discussion of Alinsky's model of organizing, I try to show that these alternative practices embody a richness of democratic thought comparable to that of progressives. And in my case study, I try to give readers a visceral sense of the ways in which these different practices of individual and collective empowerment can intertwine, conflict, and support each other in real contexts. My overall argument is that we need to find ways to bring our different practices of democracy together in contexts where their different strengths and aims can support each other.

Yet, I also try to show how difficult this can be. There may not be any simple way to seamlessly integrate the different approaches I discuss. In fact, it is likely that they will necessarily operate at different levels of any large collective undertaking. The kind of collaboration possible in small groups does not work on the scale of large "public meetings." The focus on individual actualization sought by personalists will remain in tension both with moments of mass action and with the limits on individual perspectives required by collaborative contexts. And there are likely more kinds of "democratic" practices that must be sifted into the mix.

Bringing different practices together will not be easy. But there is great potential for those groups that can. In fact, neo-Alinsky community organizing groups have already begun to do so, with their large meetings, small "issue committees," and very personal "one-on-one" interview process. Even in these arenas, however, I believe (and actually argue in my series "Core Dilemmas of Community Organizing"[1]) that we have a very long way to go.

Empowerment and Education

What, specifically, does this book imply for education, in particular? In part, I believe, it fundamentally contests our current visions of what an "empowering" classroom looks like. Our best progressive pedagogies may successfully help even working-class children operate more effectively within middle-class contexts, initiating them into key capacities for professional collaboration and dialogic interaction. But what they cannot do, if Alinsky is right, is help empower them in the poor and segregated contexts in which many of them live. This means that committed progressive educators who would like to provide their students with effective tools for collective action need to think more concretely about how to initiate students into practices like Alinsky's.

In most poor communities there is an enormous void of knowledge about collective empowerment. Inner-city youth, for example, usually know about the civil rights movement, but they almost never have any understanding of the strategies that allowed this movement to emerge and maintain itself against almost overwhelming repression. The names Martin Luther King Jr., Rosa Parks, and Cesar Chavez, and even those from local histories (e.g., Father Groppi and Vel Phillips in my city of Milwaukee) have become magical symbols lacking much coherent actionable content. Especially in America's increasingly fragmented, impoverished urban communities, connections to groups' own deep histories of collective resistance have increasingly been lost.

The "mutualism" of working-class culture survives to some extent in unions, the few deeply rooted working-class and poor communities that remain, and in the ties of extended and constructed families. But unions themselves are under constant assault from global capitalism. And the few local cultures of communal support are dissolving under the predations of a justice system that has increasingly criminalized poverty and skin color, with the high mobility that comes with modern urban poverty, and more.

As I wrote elsewhere, inner-city residents, "living in sometimes dangerous contexts, working many hours merely to sustain themselves or resigned to unemployment or underemployment, . . . retreat from common spaces and unreliable or oppressive institutions. Under such conditions, even personal networks with family and friends break down . . . Participation in community activities plummets, eliminating public spaces for dialogue and engagement. In many central-city areas, 'aside from the fact that they reside on the same streets and even live in the same apartment buildings,' residential mobility, among other issues, has led residents to see each other as strangers, perceiving 'themselves as having little if anything in common.'"[2] These communities desperately need (and increasingly lack) leaders with the basic skills and strategies for effective organizing.[3]

Outside of classrooms a movement for "youth organizing" has been emerging in communities. To Alinsky's adult model, youth organizers have added a range of social services and developmental supports in response to the often-challenging and still-forming lives of youth.[4] This work needs to be extended, and educators have much to offer if they are willing to learn about the complexities of collective action. Very little of this work, however, has made it inside schools.

This should not be a surprise. Schools, especially those populated by marginalized students, naturally recoil from the collective empowerment of youth. Students often experience school as an oppressive institution they would like to change, and school staff have little incentive to intensify this internal resistance. Schools also rightly fear attacks from outside about any pedagogy that seems to have potential for the "politicization" or "radicalization" of education.

So why bother to even explore possibilities for efforts in K–12 education related to youth organizing in school?

We must accept, I think, that only a very small number of teachers and schools will ever have much interest in pursuing a "youth-organizing" vision. But the dearth of knowledge about collective empowerment among residents of low-income communities is so severe that even small changes could make a difference. Even a small shift in the number of people who are initiated into these skills might be able to catalyze more significant changes. There is at least the hope that initiating more youth into practices like these within *and* outside of schools might create small cohorts of young adults with the capacities and inclinations to educate others.

At least a few unique "alternative" schools scattered around the nation have unique missions, governance structures, and relationships with students that could fit quite well with aspects of pedagogy related to youth organizing.[5] And, in fact, some examples do exist of teachers and schools that have tried to support student social action.[6] More educators would likely pursue such efforts if they had some sense of how to do it. With a limited research base, however, they mostly have to start from scratch.

Many unsung educators likely have tried but experienced failure and then stopped trying. As a result, I believe it is critical, especially in impoverished areas, that we begin to at least try to imagine how schools might become more empowering.

Organic Intellectuals

Gramsci argued that if different classes were to successfully respond to the larger structural challenges of society, they needed what he called "organic intellectuals" who could help them construct coherent conceptualizations of their world. Authentic organic intellectuals would act out of the actual conditions and cultural practices of those they sought to support, feeling "the elementary passions of the people, understanding them and therefore explaining and justifying them in the particular historical situation."[7] Without this depth of understanding, he worried that "relations between ... intellectual[s] and the people-nation" would be "reduced to ... relationships of a purely bureaucratic and formal order." He emphasized, therefore, that organic intellectuals must be "active in participation in practical life, as constructor, organizer, [and] 'permanent persuader,'" not ivory-tower thinkers.[8] With respect to the working class, his description fits labor leaders like John L. Lewis (who mentored Alinsky), Mother Jones, and A. Philip Randolph, as well as Alinsky.

Some middle-class university intellectuals would like to set themselves up as "organic intellectuals" for the less privileged. The problem with this aspiration is that as one moves up the scales of power within schools and universities, one increasingly takes on characteristics of the middle class. Even academics from working-class backgrounds generally achieve success precisely *because* they have learned to moderate or even suppress whatever working-class tendencies they might embody. Although there are growing efforts among working-class academics to contest the dominance of middle-class practices within the academy, the embattled tone of most of these critiques indicates how far we are from a situation in which working-class practices and dispositions are widely respected. Working-class academics often report being caught between two worlds, fully belonging to neither. In any case, as these accounts repeatedly note, the vast majority of academics come from solidly middle-class, if not upper-middle-class, backgrounds.[9]

Further, educational scholars, especially, generally only meet members of the working class in institutions like schools. Contexts like these generally place working-class people in positions of inferiority. It is not an exaggeration to say that schools are some of the most unlikely places in our society for members of the middle and working classes to meet on equal terms.[10]

Given the realities of the contexts we spend most of our time in, it seems doubtful that a significant number of middle-class scholars or educators could transform themselves into "organic intellectuals" for the working class. Only in rare cases, like that of Alinsky, who spent years of intense personal engagement in Chicago ghettos before he attempted to organize (and frankly was unique in many ways), is someone likely to be able to become an organic intellectual. Reading a lot of books will not get you there. Neither will relationships that do not (or cannot) move beyond researcher and researched, or server and served.[11]

Building Bridges?

Instead of becoming "organic intellectuals," one important first step for middle-class intellectuals may be, instead, to take on what Fred Rose called the role of "bridge builders." This role requires one to acknowledge the limitations of one's perspective and practices, to let go of the tendency to despair about the failure of "others" to become like you. At the same time, we should not take this self-critique too far. Progressives do, in fact, bring useful resources to the table with them.

These resources go beyond progressive practices of democracy. More generally, as Alinsky increasingly acknowledged later in his life, it is unlikely that poor and working-class people can generate fundamental changes in society by themselves. Only when working- and middle-class people come together in alliances are either likely to generate the kind of collective power necessary to successfully confront the status quo.

Further, members of the middle class, for good or ill, often end up working as organizers. Certainly most of the organizers that Alinsky worked with were solidly middle class. Even today, despite efforts to recruit more systematically from local communities, many organizers in community and labor organizations have similar backgrounds. And this should not be entirely surprising. Middle-class youth often have the luxury of social critique and access to more systematic educational opportunities in which such critique is encouraged. Further, as many participants in the civil rights movement acknowledged, members of the middle class often bring many key skills and are helpful when engaging and helping others to engage with their brethren in the kind of status quo organizations and government structures against which organizing groups generally struggle.

Rose has suggested that even people deeply embedded in one class or another can become bridge builders who develop relationships with individuals from a different class. As bridge builders develop trust with those from different cultures, Rose argues, "each side comes to understand the other's perspective better" and comes "to appreciate its unique contribution" to a common effort.[12] Bridge builders, in Rose's vision, do not become organic intellectuals for other groups. Instead, they gain the capacity to teach those from their own class about the strengths of others, at the same time that they educate those from other classes about the potentials resident in their own class practices.

Their insider work is perhaps most crucial because, as Rose emphasized, "people do not often alter their perception because of the ideas of those whom they consider outsiders.... A difference with a friend always remains a potential source of learning, while a difference with a stranger is readily dismissed." Because being a bridge builder often involves challenging deeply held commitments and the seemingly universal practices of one's own culture, playing this role is inevitably "fraught with difficulties as well as political risks." Through the efforts of bridge builders, however, different groups can learn to work together and also to draw on initially alien practices that may enrich their own efforts to foster social change.[13]

As a middle-class scholar myself, this book was written mostly for other insiders—for an audience of other similarly placed scholars and professionals. My goal is mainly to trace some limits of the thinking of middle-class professional scholars, educators, and others. In this way, I seek to build bridges between middle-class scholars, educators, and working-class people around approaches to empowerment. Ultimately, with Lichterman, Rose, and others, I am increasingly convinced that those of us who are interested in fostering social change and creating a more equitable and democratic society have little chance of success until more of us take upon ourselves the tensions involved in becoming bridge builders across class.[14]

Conclusion

The vision of the progressives was one of hope, of the possibility of progress in the face of seemingly intractable circumstances. And they accomplished much in America, even if democratic transformation was not one of their areas of success. Going forward, those of us who still wish for a more democratic society need to become more self critical about our deeply held but inevitably culturally rooted commitments. If there is to be any hope for redeeming the lost dreams of our progressive forbearers, middle-class scholars and educators must reach out and grasp the democratic possibilities inherent in the alien practices of working-class social struggle.

Notes

Acknowledgments

1. In many ways, Arendt and the collaborative progressives were fellow travelers, with respect to their vision of collective action. For those who are interested, my earlier essays on Arendt include Aaron Schutz, "Is Political Education an Oxymoron? Hannah Arendt's Resistance to Public Spaces in Schools," in *Philosophy of Education 2001* (Urbana: Philosophy of Education Society, 2002), 373–93; Aaron Schutz, "Theory as Performative Pedagogy: Three Masks of Hannah Arendt," *Educational Theory* 51 no. 2 (2001): 127–50; Aaron Schutz, "Teaching Freedom? Postmodern Perspectives," *Review of Educational Research* 70 no. 2 (2000): 215–51; Aaron Schutz, "Caring in Schools Is Not Enough: Community, Narrative, and the Limits of Alterity," *Educational Theory* 48, no. 3 (1998): 373–93.

Introduction

1. See Michael McGerr, *A Fierce Discontent: The Rise and Fall of the Progressive Movement in America, 1870–1920* (New York: Oxford University Press, 2005); David W. Southern, *The Progressive Era and Race: Reaction and Reform, 1900–1917* (Wheeling, IL: Harlan Davidson, 2005).
2. Gender is also an important complicating factor in this case. As Susan Stall and Randy Stoecker, among others have pointed out, democratic solidarity tends to be associated with male-dominated contexts, while collaborative and personalist democracy draws more from women's traditions of collective engagement in America. This intersection between gender and class ways of being is a fascinating issue that I will not examine in detail. See Stall and Stoecker, "Community Organizing or Organizing Community? Gender and the Crafts of Empowerment," *Gender and Society* 12, no. 6 (1998): 729–56.
3. The referendum and initiative process available in some states for the direct passage of legislation by a vote of the people is the most important exception, although recent research has shown that this process is often not very democratic in the sense the collaborative progressives meant this, especially because of the influence of special interest dollars.
4. Nel Noddings, *The Challenge to Care in Schools: An Alternative Approach to Education* (New York: Teachers College Press, 1992).

5. Lynn M. Sanders, "Against Deliberation," *Political theory* 25, no. 3 (1997): 347–76; Iris Marion Young, *Intersecting Voices: Dilemmas of Gender, Political Philosophy, and Policy* (Princeton, NJ: Princeton University Press, 1997).
6. Paul Lichterman, *The Search for Political Community: American Activists Reinventing Commitment* (New York: Cambridge University Press, 1996); Fred Rose, *Coalitions Across the Class Divide: Lessons from the Labor, Peace, and Environmental Movements* (Ithaca, NY: Cornell University Press, 2000).
7. James Paul Gee, *Social Linguistics and Literacies: Ideology in Discourses* (New York: Falmer Press, 1990); Eric H. F. Law, *The Wolf Shall Dwell with the Lamb: A Spirituality for Leadership in a Multicultural Community* (St. Louis, MO: Chalice, 1993).
8. Aaron Schutz, "Social Class and Social Action: The Middle-Class Bias of Democratic Theory in Education," *Teachers College Record* 110, no. 2 (2008): 405–48; Aaron Schutz, "John Dewey's Conundrum: Can Democratic Schools Empower?" *Teachers College Record* 103, no. 2 (2001): 267–302; Aaron Schutz, "John Dewey and 'A Paradox of Size': Democratic Faith at the Limits of Experience," *American Journal of Education* 109, no. 3 (2001): 287–319.

Chapter 1

1. This book contributes to an emerging line of work among historians of progressivism. See, for example, Michael McGerr, *A Fierce Discontent: The Rise and Fall of the Progressive Movement in America, 1870–1920* (New York: Oxford University Press, 2005), and Shelton Stromquist, *Reinventing "The People": The Progressive Movement, the Class Problem, and the Origins of Modern Liberalism* (Urbana: University of Illinois Press, 2006).
2. Jean Anyon, "Social Class and the Hidden Curriculum of Work," *Journal of Education* 162, no. 1 (1980): 67–92.
3. Joseph Kahne and Joel Westheimer, "Teaching Democracy: What Schools Need To Do," *Phi Delta Kappan* 85, no. 1 (2003): 34–40, 57–67.
4. Nel Noddings, *The Challenge to Care in Schools: An Alternative Approach to Education* (New York: Teachers College Press, 1992).
5. David F. Labaree, *The Trouble with Ed Schools* (New Haven, CT: Yale University Press, 2004).
6. Anthony Giddens, *Capitalism and Modern Social Theory: An Analysis of the Writings of Marx, Durkheim and Max Weber* (Cambridge: Cambridge University Press, 1971); Erik Olin Wright, *Approaches to Class Analysis* (New York: Cambridge University Press, 2005).
7. Actually, he argued that cultural capital can be transformed, in some cases, into material capital and that in most cases, material capital is dominant. And, of course, Marx and others also influenced Bourdieu.
8. David Swartz, *Culture and Power: The Sociology of Pierre Bourdieu* (Chicago: University of Chicago Press, 1998), 109. Swartz argued that Bourdieu's work had "little to say about what *collective* forms of class struggle look like" (187). And his vision of working-class culture often seems quite limited, focusing on how

it is "highly constrained by primary necessities" (176). I am less conversant with more recent publications which may temper this pattern.
9. Stuart M. Blumin, *The Emergence of the Middle Class: Social Experience in the American City, 1760–1900* (Cambridge: Cambridge University Press, 1989), 434.
10. Pierre Bourdieu, *Distinction: A Social Critique of the Judgment of Taste* (Cambridge, MA: Harvard University Press, 1984).
11. In fact, Bourdieu was uncomfortable with "single-dimensional scales and cumulative indices that locate individuals and groups by position in social structure," preferring "multidimensional analysis." Swartz, *Culture and Power*, 129.
12. See Max Weber, *Economy and Society* (Berkeley: University of California Press, 2002); Max Weber, *The Methodology of the Social Sciences*, trans. Edward Schils (Glencoe, IL: Free Press, 1949); Martin Albrow, *Max Weber's Construction of Social Theory* (New York: St. Martin's Press, 1990); Susan J. Hekman, *Weber, the Ideal Type, and Contemporary Social Theory* (Notre Dame, IN: University of Notre Dame Press, 1983).
13. Alvin Ward Gouldner, *The Future of Intellectuals and the Rise of the New Class* (New York: Seabury Press, 1979), 8. Note that Gouldner, at the end of his career, made an argument about the relationship between Marx's theory and class background very similar to the one I am making about Dewey and the progressives in this volume, although I did not realize this until late in my writing process. Gouldner's admittedly idiosyncratic writings will likely become more important if, as seems likely to me, Marxian theory returns to prominence in academic thought in the humanities and social sciences.
14. See Andrew Milner, *Class* (Thousand Oaks, CA: Sage, 1999).
15. I have selected here a fairly narrow understanding of the rich complexity of Bourdieu's concept of the "field," a choice that he often made in his own work as well (see Swartz, *Culture and Power*, 117–42).
16. See Alan Trachtenberg, *The Incorporation of America: Culture and Society in the Gilded Age* (New York: Hill and Wang, 2007).
17. Michèle Lamont, *Money, Morals, and Manners: The Culture of the French and American Upper-Middle Class* (Chicago: University of Chicago Press, 1992).
18. Rick Fantasia, *Cultures of Solidarity: Consciousness, Action, and Contemporary American Workers* (Berkeley: University of California Press, 1988); Alfred Lubrano, *Limbo: Blue-Collar Roots, White-Collar Dreams* (Hoboken, NJ: Wiley, 2004); Betsy Leondar-Wright, *Class Matters: Cross-Class Alliance Building for Middle-Class Activists* (New York: New Society Publishers, 2005). My analysis is indebted to prior writings on education and social class. Classic works by Michael Apple, Jean Anyon, Samuel Bowles and Herbert Gintes, Martin Carnoy and Henry Levin, and Henry Giroux, for example, informed my general thinking. It is important to acknowledge, however, that in the past few decades a focus on social class in the education literature largely disappeared in favor of a broad range of discussions of postmodernism. A few education scholars, including Apple and Richard Brosio, fought with limited success to maintain

and extend our understandings of social class during this fallow period. More recently, however, questions of social class seem to be returning to prominence, as evidenced by a range of attacks on postmodernism from a Marxian perspective. Contemporary scholars like Ellen Brantlinger and Annette Lareau have also conducted powerful empirical analyses of the effects of class culture on schools and family life. All this work influenced my efforts to understand how class-based practices might inform educational scholars. See Anyon, "Hidden Curriculum," 67–92; Michael W. Apple, *Ideology and Curriculum* (London: Routledge and K. Paul, 1979); Samuel Bowles and Herbert Gintis, *Schooling in Capitalist America: Educational Reform and the Contradictions of Economic Life* (New York: Basic Books, 1976); Ellen A. Brantlinger, *Dividing Classes: How the Middle Class Negotiates and Rationalizes School Advantage* (New York: Routledge Falmer, 2003); Richard A. Brosio, *A Radical Democratic Critique of Capitalist Education* (New York: P. Lang, 1994); Mike Cole et al., *Red Chalk: On Schooling, Capitalism and Politics: Mike Cole, Dave Hill and Glenn Rikowski in Discussion with Peter McLaren* (Brighton: The Institute for Education Policy Studies, 2001); Martin Carnoy and Henry M. Levin, *Schooling and Work in the Democratic State* (Stanford, CA: Stanford University Press, 1985); Henry A. Giroux, *Theory and Resistance in Education: A Pedagogy for the Opposition* (South Hadley, MA: Bergin and Garvey, 1983); Annette Lareau, *Unequal Childhoods: Class, Race, and Family Life* (Berkeley: University of California Press, 2003); Cameron McCarthy and Michael Apple, "Race, Class, and Gender in American Educational Research: Toward a Nonsynchronous, Parallelist Position," in *Race, Class, and Gender in American Education*, ed. Lois Weis (Albany, NY: SUNY Press, 1988), 9–25.

19. Blumin, *Emergence of the Middle Class*; Thomas R. Mahoney, "Middle-Class Experience in the United States in the Gilded Age, 1865–1900," *Journal of Urban History* 31 (2005): 356–66.
20. See Susan Porter Benson, *Counter Cultures: Saleswomen, Managers, and Customers in American Department Stores, 1890–1940* (Urbana: University of Illinois Press, 1986); Trachtenberg, *Incorporation of America*.
21. Blumin, *Emergence of the Middle Class*, 233.
22. Mahoney, "Middle-Class Experience," 361, 360.
23. Ibid., 363, 361, 363. See Rosabeth Moss Kanter, *Men and Women of the Corporation* (New York: Basic Books, 1977).
24. Blumin, *Emergence of the Middle Class*, 187.
25. Ibid.
26. "Surveys show that two out of three middle- and upper-class high school graduates attended a four-year college, as compared to just one of five from the working and lower classes." Lubrano, *Limbo*, 11. This statistic would certainly become even more stark if the relative quality and reputation of the colleges attended were taken into account.
27. Robert Kanigel, *The One Best Way: Frederick Winslow Taylor and the Enigma of Efficiency* (New York: Viking, 1997), 538.

28. Susan Curtis, *A Consuming Faith: The Social Gospel and Modern American Culture* (Baltimore: Johns Hopkins University Press, 1991), 24.
29. Trachtenberg, *Incorporation of America*, 87.
30. Ibid., 87, 88.
31. Harry Braverman, *Labor and Monopoly Capital: The Degradation of Work in the Twentieth Century* (New York: Monthly Review Press, 1975), 66.
32. Joseph George Rayback, *A History of American Labor* (New York: Free Press, 1966).
33. Trachtenberg, *Incorporation of America*.
34. Rayback, *American Labor*.
35. Trachtenberg, *Incorporation of America*, 90.
36. Mike Davis, *Prisoners of the American Dream: Politics and Economy in the History of the U.S. Working Class* (London: Verso, 1986), 19.
37. See ibid.; Rayback, *American Labor*.
38. See McGerr, *Fierce Discontent*; David Montgomery, *The Fall of the House of Labor: The Workplace, the State, and American Labor Activism, 1865–1925* (New York: Cambridge University Press, 1987).
39. Blumin, *Emergence of the Middle Class*, 233.
40. Montgomery, *House of Labor*, 2.
41. Trachtenberg, *Incorporation of America*, 88.
42. Stansell, cited in Blumin, *Emergence of the Middle Class*, 189.
43. McGerr, *Fierce Discontent*, 18.
44. Braverman, *Labor and Monopoly Capital*.
45. Benson, *Counter Cultures*; Kanter, *Men and Women of the Corporation*; Robin Leidner, *Fast Food, Fast Talk: Service Work and the Routinization of Everyday Life* (Berkeley: University of California Press, 1993).
46. Despite critiques of Braverman's thesis, there seems to be a general agreement that deskilling remains at least a "major tendential presence within the development of the capitalist labor process" that disproportionately affects those on the bottom. Peter Meiksins, "Labor and Monopoly Capital for the 1990s: A Review and Critique of the Labor Process Debate," *Monthly Review* 46, no. 6 (1994), 5; see James Paul Gee, Glynda A. Hull, and Colin Lankshear, *The New Work Order: Behind the Language of the New Capitalism* (Boulder, CO: Westview, 1996); Montgomery, *House of Labor*.
47. Ken Estey, *A New Protestant Labor Ethic at Work* (Cleveland, OH: Pilgrim, 2002); Gee, Hull, and Lankshear, *New Work Order*.
48. David K. Brown, *Degrees of Control: A Sociology of Educational Expansion and Occupational Credentialism* (New York: Teachers College Press, 1995), 56.
49. Gee, Hull, and Lankshear, *New Work Order*.
50. Aaron Schutz, "Home Is a Prison in the Global City: The Tragic Failure of School-Based Community Engagement Strategies," *Review of Educational Research* 76, no. 4 (2006): 691–743.
51. For recent examples, see Betty Hart and Todd R. Risley, *Meaningful Differences in the Everyday Experiences of Young American Children* (Baltimore, MD: Brooks, 1995); Lareau, *Unequal Childhoods*; Jonathan R. H. Tudge et

al., "Parent's Child-Rearing Values and Beliefs in the United States and Russia: The Impact of Culture and Social Class," *Infant and Child Development* 9 (2000): 105–22.
52. Annette Lareau, "Invisible Inequality: Social Class and Childrearing in Black Families and White Families," *American Sociological Review* 67 (2002): 747.
53. Basil B. Bernstein, *Class, Codes and Control* (London: Routledge, 1971), 171.
54. Tudge et al., "Parent's Child-Rearing Values," 107.
55. Hart and Risley, *Meaningful Differences*, 133.
56. See Anyon, "Hidden Curriculum"; Hart and Risley, *Meaningful Differences*; Melvin L. Kohn, and Carmi Schooler, *Work and Personality: An Inquiry into the Impact of Social Stratification* (Norwood, NJ: Ablex, 1983).
57. Lareau, "Social Class and the Daily Lives of Children," *Childhood* 7, no. 2 (2000): 161.
58. Lareau, *Unequal Childhoods*.
59. Leondar-Wright, *Class Matters*.
60. Lareau, *Unequal Childhoods*.
61. Bernstein, *Class, Codes and Control*; Brown, *Degrees of Control*; Gee, Hull, and Lankshear, *New Work Order*; Trutz von Trotha and Richard Harvey Brown, "Sociolinguistics and the Politics of Helping," *Acta Sociologica* 25, no. 4 (1982): 373–87. Bernstein famously argued that the conditions of working-class life have produced a "restricted" discursive "code" that assumes "that speaker and hearer share a common frame of reference." Bernstein, *Class, Codes and Control*, 119. He claimed that because the restricted code is less explicit about the assumptions that lie behind particular statements, it is less conducive to the kind of abstract discourse and thought prominent in middle-class settings. This argument has frequently been attacked by those who perceived an implicit denigration of working-class thought, even though that is not what he had intended. I think Von Trotha and Brown's analysis of the different (but equally demanding) cognitive demands in working-class and middle-class settings is a better way to frame Bernstein's ideas.
62. Von Trotha and Brown, "Politics of Helping," 383.
63. Michèle Lamont, *The Dignity of Working Men: Morality and the Boundaries of Race, Class, and Immigration* (Cambridge, MA: Harvard University Press, 2000), 36.
64. Von Trotha and Brown, "Politics of Helping," 383.
65. Lamont, *Money, Morals, and Manners*, 39.
66. Brown, *Degrees of Control*, 56, 61, 62.
67. Ernest T. Pascarella and Patrick T. Terenzini, *How College Affects Students: A Third Decade of Research* (San Francisco, CA: Jossey-Bass, 2005).
68. Brown, *Degrees of Control*; Pascarella and Terenzini, *How College Affects Students*.
69. Lubrano, *Limbo*; Michelle M. Tokarczyk and Elizabeth A. Fay, *Working Class Women in the Academy: Laborers in the Knowledge Factory* (Amherst: University of Massachusetts Press, 1993).
70. Lamont, *Money, Morals, and Manners*, 33.

71. Lareau, "Invisible Inequality," 763.
72. Godfrey J. Ellis and Larry R. Peterson, "Socialization Values and Parental Control Techniques: A Cross-Cultural Analysis of Child-Rearing," *Journal of Comparative Family Studies* 23 (1992): 39–45; Kohn and Schooler, *Work and Personality*.
73. Lareau, "Invisible Inequality," 747.
74. James Paul Gee, "Teenagers in New Times: A New Literacy Studies Perspective," *Journal of Adolescent and Adult Literacy* 43, no. 5 (2000): 412–20; Shirley Brice Heath, *Ways With Words: Language, Life and Work in Communities and Classrooms* (New York: Cambridge University Press, 1983).
75. Peggy J. Miller, Grace E. Cho, and Jeana R. Bracey, "Working-Class Children's Experience through the Prism of Personal Storytelling," *Human Development* 48, no. 3 (2005): 131.
76. Leondar-Wright, *Class Matters*, 22.
77. Lubrano, *Limbo*, 20.
78. Zygmunt Bauman, *City of Fears, City of Hopes* (London: Goldsmiths College, 2003), 16–17; also see Schutz, "Home Is a Prison."
79. For example, Kohn and Schooler, *Work and Personality*.
80. Glynda Hull, "Critical Literacy and Beyond: Lessons Learned from Students and Workers in a Vocational Program and on the Job," *Anthropology and Education Quarterly* 24, no. 4 (1993): 373–96; Leidner, *Fast Food, Fast Talk*; Katherine S. Newman, *No Shame in My Game: The Working Poor in the Inner City* (New York: Knopf, 1999).
81. Von Trotha and Brown, "Politics of Helping," 380–82.
82. In fact, most of the technological advancements of the industrial revolution— the invention of the steam engine, and so forth—resulted from the experimentation and pragmatic adjustments of those who, today, would be classified as working-class mechanics. See Celeste Connor, *Democratic Visions: Art and Theory of the Stieglitz Circle, 1924–1934* (Berkeley: University of California Press, 2001); Peter T. Manicas, *A History and Philosophy of the Social Sciences* (New York: B. Blackwell, 1988).
83. Fred Rose, *Coalitions Across the Class Divide: Lessons from the Labor, Peace, and Environmental Movements* (Ithaca: Cornell University Press, 2000), 24.
84. Bernstein, *Class, Codes and Control*, 9, 176; see also Patricia Hill Collins, *Black Feminist Thought: Knowledge, Consciousness, and the Politics of Empowerment* (New York: Allen and Unwin, 1990).
85. Aaron Schutz, "Rethinking Domination and Resistance: Challenging Postmodernism," *Educational Researcher* 33, no. 1 (2004): 15–23.
86. David W. Livingstone and Peter H. Sawchuk, "Hidden Knowledge: Working-Class Capacity in the 'Knowledge-Based Economy,'" *Studies in the Education of Adults* 37, no. 2 (2005): 112.
87. Thomas J. Gorman, "Social Class and Parental Attitudes toward Education," *Journal of Contemporary Ethnography* 27, no. 1 (1998): 115.
88. Benson, *Counter Cultures*; also see Paul Willis, *Learning to Labour: How Working-Class Kids Get Working-Class Jobs* (New York: Columbia University Press, 1982).

89. Thomas J. Gorman, "Social Class and Parental Attitudes Toward Education," *Journal of Contemporary Ethnography* 27, no. 1 (1998): 10–44; Richard Sennett and Jonathan Cobb, *The Hidden Injuries of Class* (New York: Knopf, 1972); Bernice Lott, "Cognitive and Behavioral Distancing from the Poor," *American Psychologist* 57, no. 2 (2002): 100–110.
90. Gorman, "Parental Attitudes," 101. See Brantlinger, *Dividing Classes*; Lott, "Distancing from the Poor"; Sennett and Cobb, *Hidden Injuries*.
91. Lamont, *Dignity of Working Men*.
92. Lubrano, *Limbo*, 10. See Rose, *Class Divide*.
93. I use this phrase differently than Durkheim, who, in a simple sense, was referring more broadly to the distinction between premodern and modern societies, using *organic* to describe the individualism, division of labor, and complex interdependence of modern society.
94. Von Trotha and Brown, "Politics of Helping," 381.
95. Collins, *Black Feminist Thought*.
96. Lareau, *Unequal Childhoods*.
97. Gorman, "Parental Attitudes," 106; see also Sennett and Cobb, *Hidden Injuries*.
98. Annette Lareau and Wesley Shumar, The Problem of Individualism in Family-School Policies. *Sociology of Education* 69 (1996): 30.
99. Anyon, "Hidden Curriculum."
100. James Paul Gee, *Social Linguistics and Literacies* (New York: Falmer, 1990).
101. Bernstein, *Class, Codes and Control*; Lisa Delpit, *Other People's Children: Cultural Conflict in the Classroom* (New York: New Press, 1995); Heath, *Ways With Words*.
102. Helen Lucey, June Melody, and Valerie Walkerdine, "Uneasy Hybrids: Psychosocial Aspects of Becoming Educationally Successful for Working-Class Young Women," *Gender and Education* 15, no. 3 (2003): 285.
103. Peter Kaufman, "Learning to Not Labor: How Working-Class Individuals Construct Middle-Class Identities," *Sociological Quarterly* 44 (2003): 481–504.
104. Lucey, Melody, and Walkerdine, "Uneasy Hybrids," 296.
105. Lubrano, *Limbo*.
106. Gorman, "Parental Attitudes," 25.
107. Lubrano, *Limbo*, 32, 2.
108. Lamont, *Dignity of Working Men*, 30; see also Lubrano, *Limbo*.
109. Thomas J. Gorman, "Cross-Class Perceptions," *Sociological Spectrum* 20, no. 1 (2000): 104.
110. Trachtenberg, *Incorporation of America*, 40.
111. Rayback, *A History of American Labor*, 168.
112. See Curtis, *A Consuming Faith*; McGerr, *A Fierce Discontent*.
113. McGerr, *Fierce Discontent*; Stromquist, *Reinventing "The People."*
114. See, for example, Barbara Eherenreich and John Erenreich, "The Professional-Managerial Class," in *Between Labor and Capital*, ed. Pat Walker (Boston: South End Press, 1979), 5–45. Lamont, in *Money, Morals, and Manners*, referred to a related contrast between "for-profit" and "nonprofit" employees.

115. Robert H. Wiebe, *The Search for Order, 1877–1920* (New York: Hill and Wang, 1966), 154, 153, 176.
116. See, for example, Frederick Winslow Taylor, *The Principles of Scientific Management* (New York: Norton, 1967).
117. See Kanigel, *The One Best Way*, for an overview.
118. Wiebe, *Search for Order*, 152, 169.
119. Walter Lippmann, *Public Opinion* (New York: Harcourt, Brace, 1922).
120. Ibid.; Robert B. Westbrook, *John Dewey and American Democracy* (Ithaca, NY: Cornell University Press, 1991).
121. James C. Scott, *Seeing Like a State: How Certain Schemes to Improve the Human Condition Have Failed* (New Haven, CT: Yale University Press, 1998).
122. Trachtenberg, *Incorporation of America*.
123. McGerr, *Fierce Discontent*, 65, 69.
124. Curtis, *Social Gospel*; James T. Kloppenberg, *Uncertain Victory: Social Democracy and Progressivism in European and American Thought, 1870–1920* (New York: Oxford University Press, 1986); McGerr, *Fierce Discontent*; Stromquist, *Reinventing "The People."*
125. John Dewey, *Democracy and Education: An Introduction to the Philosophy of Education* (New York: Macmillan, 1916).
126. Katherine Camp Mayhew and Anna Camp Edwards, *The Dewey School: The Laboratory School of the University of Chicago, 1896–1903* (New York: D. Appleton-Century, 1936), 79, 339.
127. Cornel West, *The American Evasion of Philosophy: A Genealogy of Pragmatism* (Madison: University of Wisconsin Press, 1989), 76; see also Walter Feinberg, *Reason and Rhetoric: The Intellectual Foundations of 20th-Century Liberal Educational Policy* (New York: Wiley, 1975).
128. See C. A. Bowers, *The Progressive Educator and the Depression: The Radical Years* (New York: Random House, 1969); George S. Counts, *Dare the School to Build a New Social Order?* (New York: John Day, 1967); Feinberg, *Reason and Rhetoric*; Peter S. Hlebowitsh and William G. Wraga, "Social Class Analysis in the Early Progressive Tradition," *Curriculum Inquiry* 25, no. 1 (1995): 7–22. In fact, one might argue that there was a veiled paternalism in the writings of Counts and later social reconstructionists when they assumed that the traditions of the working class needed to be altered by middle-class educators.
129. Curtis, *Social Gospel*; Kloppenberg, *Uncertain Victory*; McGerr, *Fierce Discontent*.
130. Wiebe, *Search for Order*. Dewey became increasingly unconvinced about this position in his later years.
131. McGerr, *Fierce Discontent*, 135, 134, italics added.
132. Jane Addams, *The Jane Addams Reader*, ed. Jean Bethke Elshtain (New York: Basic Books, 2002), 488.
133. Jean Bethke Elshtain, *Jane Addams and the Dream of American Democracy: A Life* (New York: Basic Books, 2002), 124.
134. Feinberg, *Reason and Rhetoric*; Kloppenberg, *Uncertain Victory*; McGerr, *Fierce Discontent*.

135. Montgomery, *House of Labor*.
136. Mother Jones, *Mother Jones Speaks*, ed. Phillip Foner (New York: Monad Press, 1983), 147.
137. See Stuart D. Brandes, *American Welfare Capitalism, 1880–1940* (Chicago: University of Chicago Press, 1976); Fantasia, *Cultures of Solidarity*; Montgomery, *House of Labor*.
138. Estey, *Protestant Labor Ethic at Work*; James Reinhart, "Transcending Taylorism and Fordism?" in *The Critical Study of Work*, ed. Rick Baldoz, Charles Kroeber, and Philip Kraft (Philadelphia, PA: Temple University Press, 2001), 179–93.
139. For the sake of brevity and simplicity, going forward I will refer to these two eras as the 1920s and the 1960s except in cases where this might be confusing.
140. As other scholars have noted, the transcendentalists deeply influenced collaborative democrats like Dewey as well. See David A. Granger, *John Dewey, Robert Pirsig, and the Art of Living: Revisioning Aesthetic Education* (New York: Palgrave Macmillan, 2006); Naoko Saito, *The Gleam of Light: Moral Perfectionism and Education in Dewey and Emerson* (New York: Fordham University Press, 2005).
141. The most important work on the 1920s personalists is Casey N. Blake, *Beloved Community: The Cultural Criticism of Randolph Bourne, Van Wyck Brooks, Waldo Frank and Lewis Mumford* (Chapel Hill, NC: University of North Carolina Press, 1990); the best book for understanding the personalist perspective of the 1960s is Paul Goodman, *Growing Up Absurd* (New York: Vintage, 1962); and the best book on the free schools movement is Ron Miller, *Free Schools, Free People: Education and Democracy After the 1960s* (Albany: State University of New York Press, 2002).
142. William E. Leuchtenberg, *Perils of Prosperity, 1914–1932* (Chicago: University of Chicago Press, 1993), 151.
143. Goodman, *Growing Up Absurd*, 129.
144. Ibid., 80.
145. See Margaret Naumburg, *The Child and the World: Dialogues in Modern Education* (New York: Harcourt, Brace and Co., 1928); Caroline Pratt and Jessie Stanton, *Before Books* (New York: Adelphi, 1926); A. S. Neill, *Summerhill: A Radical Approach to Child Rearing* (New York: Hart, 1960).
146. Linda Jill Markowitz, *Worker Activism After Successful Union Organizing* (Armonk, NY: M. E. Sharpe, 2000); Judith Stepan-Norris, "The Making of Union Democracy," *Social Forces* 76, no. 2 (1997): 475–510.
147. Much of my understanding of community organizing is drawn not from books but from my dialogues with organizers and work with congregational organizing groups.
148. Donald C. Reitzes and Dietrich C. Reitzes, *The Alinsky Legacy: Alive and Kicking* (Greenwich, CT: JAI Press, 1987), 36.
149. Saul Alinsky, *Reveille for Radicals* (New York: Vintage, 1969), 133–34.
150. Reitzes and Reitzes, *Alinsky Legacy*, 35.
151. Alinsky, *Reveille for Radicals*, 130.

152. Mark R. Warren, *Dry Bones Rattling: Community Building to Revitalize American Democracy* (Princeton, NJ: Princeton University Press, 2001), 336.
153. Ibid., 64.
154. Edward T. D. Chambers and Michael A. Cowan, *Roots for Radicals: Organizing for Power, Action, and Justice* (New York: Continuum, 2003).
155. See Warren, *Dry Bones Rattling*.
156. Lichterman, *The Search for Political Community: American Activists Reinventing Community* (New York: Cambridge University Press, 1996); Rose, *Coalitions*. Also see David Croteau, *Politics and the Class Divide: Working People and the Middle-Class Left* (Philadelphia: Temple University Press, 1995); Linda Stout, *Bridging the Class Divide and Other Lessons for Grassroots Organizing* (Boston: Beacon, 1996); Leondar-Wright, *Class Matters*; Stephen Hart, *Cultural Dilemmas of Progressive Politics: Styles of Engagement Among Grassroots Activists* (Chicago: University of Chicago Press, 2001); Eric H. F. Law, *The Wolf Shall Dwell with the Lamb: A Spirituality for Leadership in a Multicultural Community* (Atlanta: Chalice, 1993).
157. Rose, *Coalitions*, 65.
158. Stout, *Class Divide*, 135.
159. Rose, *Coalitions*, 20.
160. Stout, *Class Divide*, 128, 20.
161. Rose, *Coalitions*, 66–67.
162. Lichterman, *Political Community*, 24.
163. Rose, *Coalitions*, 65–67.
164. Ibid., 59.
165. Ibid., 57.
166. Ibid., 58–59.
167. Ibid., 73, 63.
168. Ibid., 18.
169. Also see Dennis Shirley, *Community Organizing for Urban School Reform* (Austin: University of Texas Press, 1997); and Warren, *Dry Bones Rattling*.
170. See my series on "Core Dilemmas of Community Organizing," which appears on the Open Left blog at http://www.educationaction.org/core-dilemmas-of-community-organizing.html.
171. Law, *Wolf Shall Dwell with the Lamb*.
172. Rose, *Coalitions*, 73, 209.
173. Ibid., 26, 27. Of course, this has not always been the case, especially in the decades around the turn of the twentieth century.
174. Laurence R. Veysey, *The Emergence of the American University* (Chicago: University of Chicago Press, 1965).
175. Dorothy Ross, *The Origins of American Social Science* (Cambridge: Cambridge University Press, 1991), xiii.
176. John R. Commons, "Discussion of the President's Address," *Publications of the American Economic Association* 3, no. 1 (1890), 62–88, 287–88; in education see Counts, *Dare the Schools*.

177. Brosio, *Radical Democratic Education*; Aaron Schutz, "Teaching Freedom? Postmodern Perspectives," *Review of Educational Research* 70, no. 2 (2000): 215–51; Schutz, "Rethinking Domination and Resistance."
178. Christine A. Ogren, *The American State Normal School: An Instrument of Great Good* (New York: Palgrave Macmillan, 2005), 68.
179. David B. Tyack, *The One Best System: A History of American Urban Education* (Cambridge, MA: Harvard University Press, 1974), 72; see also Ellen Condliffe Lagemann, *An Elusive Science: The Troubling History of Education Research* (Chicago: University of Chicago Press, 2000).
180. Tyack, *One Best System*; Herbert M. Kliebard, *The Struggle for the American Curriculum: 1893–1958* (New York: Routledge, 1995).
181. Labaree, *Trouble with Ed Schools*, 156, 142.
182. Ogren, *American State Normal School*.
183. Labaree, *Trouble with Ed Schools*, 143.
184. See Michelle Fine, Lois Weis, and Linda C. Powell, "Communities of Difference: A Critical Look at Desegregated Spaces Created for and by Youth," *Harvard Educational Review* 67, no. 2 (1997): 247–61; James W. Fraser, *Reading, Writing, and Justice: School Reform as if Democracy Matters* (Albany: State University of New York Press, 1997); David T. Sehr, *Education for Public Democracy* (Albany: State University of New York Press, 1997), to name a few. The point is not that no thoughtful (as opposed to hatchet job) critiques of Deweyan democracy exist. Instead, the problem is that his general vision is so deeply embedded in the psyches of educational scholars that alternatives rarely emerge.
185. This invisibility of working-class models of democratic engagement is almost certainly magnified by the fact that, according to Bourdieu, "working-class lifestyles . . . serve as a negative reference point for the dominant class." He argued, for example, that "perhaps their sole function in the system of aesthetic positions is to serve as a foil, a negative reference point." Swartz, *Culture and Power*, 168; the latter quote is from Bourdieu.
186. Michael W. Apple and James A. Beane, *Democratic Schools* (Alexandria, VA: Association for Supervision and Curriculum Development, 1995). See Brosio, *Radical Democratic Education*.
187. Apple and Beane, *Democratic Schools*, 16, 96.
188. See Collins, *Black Feminist Thought*; William F. Tate and Gloria Ladson-Billings, "Critical Race Theory and Education: History, Theory, and Implications," *Review of Research in Education* 22 (1997): 195–250.
189. Kathleen Knight Abowitz, "Getting Beyond Familiar Myths: Discourses of Service Learning and Critical Pedagogy," *Review of Education, Pedagogy and Cultural Studies* 21, no. 1 (1999): 63–77; Jane Mansbridge et al., "Norms of Deliberation: An Inductive Study," *Journal of Public Deliberation* 2, no. 1 (2006): 1–47; Lynn M. Sanders, "Against Deliberation," *Political Theory* 25, no. 3 (1997): 347–76.

Chapter 2

1. John Dewey, cited in Jay Martin, *The Education of John Dewey: A Biography* (New York: Columbia University Press, 2002), xi.
2. James A. Good, *A Search for Unity in Diversity: The "Permanent Hegelian Deposit" in the Philosophy of John Dewey* (Lanham, MD: Lexington Books, 2006), 101.
3. Cited in Martin, *Education of John Dewey*, 32.
4. Robert Westbrook, *John Dewey and American Democracy* (Ithaca: Cornell University Press, 1990), 4.
5. Alan Ryan, *John Dewey and the High Tide of American Liberalism* (New York: W. W. Norton, 1995), 36.
6. Jean Bethke Elshtain, *Jane Addams and the Dream of American Democracy: A Life* (New York: Basic Books, 2002).
7. Laurel Tanner, *Dewey's Laboratory School: Lessons for Today* (New York: Teachers College Press, 1997), 200.
8. John Dewey, "Human Nature and Conduct," in *The Middle Works, 1899–1924: Vol. 14, 1922*, ed. Jo Ann Boydston (Carbondale: Southern Illinois University Press, 1976), 70, 112.
9. Dewey, *Democracy and Education*, 297.
10. John Dewey, *The Public and Its Problems* (Columbus, OH: Swallow Press, 1954), 188.
11. John Dewey, *Democracy and Education: An Introduction to the Philosophy of Education* (New York: Macmillan, 1916), 18.
12. John Dewey, "Time and Individuality," in *The Later Works 14: 1925–1953: Vol. 14, 1939–1941*, ed. Jo Ann Boydston (Carbondale: Southern Illinois University Press, 1988), 112.
13. Ibid., 226.
14. Jim Garrison, "John Dewey's Theory of Practical Reasoning," *Educational Philosophy and Theory* 31, no. 3 (1999): 291–312.
15. Ibid., 83, 297.
16. Katherine Camp Mayhew and Anna Camp Edwards, *The Dewey School: The Laboratory School of the University of Chicago 1896–1903* (New York: D. Appleton-Century, 1936), 300, 306, 406.
17. Ibid., 63, 358, 71, 79.
18. Ibid., 232–33.
19. Ibid., 204.
20. Ibid., 98, 339, italics added.
21. John Dewey, "School and Society," in *The Middle Works, 1899–1924: Vol. 1, 1899–1901*, ed. Jo Ann Boydston (Carbondale: Southern Illinois University Press, 1981), 104.
22. Dewey, *Democracy and Education*, 215–16.
23. Mayhew and Edwards, *Dewey School*, 314.
24. Dewey, cited in Mayhew and Edwards, *Dewey School*, 473.

25. Laura Runyon, *The Teaching of Elementary History in the Dewey School* (Chicago: University of Chicago, 1906).
26. John Dewey, "The Aim of History in Elementary Education," *Elementary School Record* 1, no. 4 (1900): 203.
27. Ibid., 201; Runyon, *Elementary History*, 54.
28. Mayhew and Edwards, *Dewey School*, 106,107.
29. C. Wright Mills, *Sociology and Pragmatism: The Higher Learning in America* (New York: Paine-Whitman Publishers, 1964), 378.
30. Dewey, *Public*, 152.
31. John Dewey, *Art as Experience* (New York: Minton, Balch and Company, 1934), 29.
32. Dewey, *Democracy and Education*, 226.
33. Ibid., 189, 220.
34. Dewey, *Art as Experience*, 108, 244, 108, italics added.
35. Steven C. Rockefeller, *John Dewey: Religious Faith and Democratic Humanism* (New York: Columbia University Press, 1991), 539.
36. Spencer J. Maxcy, "Ethnic Pluralism, Cultural Pluralism, and John Dewey's Program of Cultural Reform: A Response to Eisle," *Educational Theory* 34, no. 3 (1984): 301–5.
37. Michael McGerr, *A Fierce Discontent: The Rise and Fall of the Progressive Movement in America, 1870–1920* (New York: Oxford University Press, 2005), 189. See also Shelton Stromquist, *Reinventing "The People": The Progressive Movement, the Class Problem, and the Origins of Modern Liberalism* (Urbana: University of Illinois Press, 2006); Susan Curtis, *A Consuming Faith: The Social Gospel and Modern American Culture* (Baltimore: Johns Hopkins University Press, 1991); David W. Southern, *The Progressive Era and Race: Reaction and Reform, 1900–1917* (Wheeling, IL: Harlan Davidson, 2005).
38. Dewey believed that different cultural groups could serve different needs and welcomed the contributions that cultural diversity could make to a democratic society. But, as Maxcy in "Ethnic Pluralism" has shown, he did not approve of groups that sought to isolate themselves from the larger whole.
39. Mills, cited in Cornel West, *The American Evasion of Philosophy: A Genealogy of Pragmatism* (Madison: University of Wisconsin Press, 1989), 126.
40. Emily Robertson, "Is Dewey's Educational Vision Still Viable?" *Review of Research in Education* 18, no. 1 (1992): 335–81, notes a few examples where Dewey appears to leave open the possibility of social conflict. Also see Dewey, "The Teacher and the Public," in *The Later Works: 1925–1953: Vol. 11, 1935–1937*, ed. Jo Ann Boydston (Carbondale: Southern Illinois University Press, 1987), 161, where he argues that teachers and others should join together against "their common foe." These are among only a very small number of exceptions.
41. John Dewey, "Nationalizing Education," in *The Middle Works: 1899–1925, Vol. 10, 1916–1917*, ed. Jo Ann Boydston (Carbondale: Southern Illinois University Press, 1980), 207.
42. West, *Evasion of Philosophy*.

43. For example, Ryan, *High Tide*; Westbrook, *Dewey*.
44. Ryan, *High Tide*, 245, 295; see also, West, *Evasion of Philosophy*, 102.
45. Westbrook, *Dewey*.
46. Mayhew and Edwards, *Dewey School*, 439; Westbrook, *Dewey*, 111.

Chapter 3

1. Jane Mansbridge, "A Paradox of Size," in *From the Ground Up*, ed. George Bonnello (Boston, MA: South End Press, 1992), 159–76.
2. For discussions of this problem, see Robert Alan Dahl and Edward R. Tufte, *Size and Democracy* (Stanford, CA: Stanford University Press, 1973); Robert Alan Dahl, *On Democracy* (New Haven, CT: Yale University Press, 1998); Norberto Bobbio and Richard Bellamy, *The Future of Democracy* (Minneapolis: University of Minnesota Press, 1987); Alexis de Tocqueville, J. P. Mayer, and Max Lerner, *Democracy in America* (New York: Harper & Row, 1966). This paradox is generally "solved" through a strategy of representative government (see Dahl and Tufte, *Size and Democracy*), an option I explore later.
3. For a more recent discussion of similar challenges, see Danilo Zolo, *Democracy and Complexity* (University Park: Pennsylvania State University Press, 1992).
4. John Dewey, "Review of *Public Opinion* by Walter Lippmann," in *The Middle Works: 1899–1924: Vol. 13, 1921–1922*, ed. Jo Ann Boydston (Carbondale: Southern Illinois University Press, 1983), 344.
5. John Dewey, *The Public and Its Problems* (Columbus, OH: Swallow Press, 1954), 128.
6. David Harvey, *The Condition of Postmodernity: An Enquiry into the Origins of Cultural Change* (Cambridge, MA: Blackwell, 1989); Ulrich Beck, *World Risk Society* (Malden, MA: Polity, 1999).
7. Dewey, *Public*, 127.
8. Ibid., 58.
9. Ibid., 59.
10. Ibid., 155, 61.
11. Ibid., 77.
12. Dewey noted that *every* citizen of a public, in their official functions as voters, for example, is supposed to act as an official. This does not change the general tenor of my argument, however.
13. Ernesto Laclau, *Emancipations* (New York: Verso, 1996), 98–99.
14. Dewey, *Public*, 183.
15. Ibid., 183, 207.
16. Ibid., 210, 209.
17. Ibid., 188, italics added.
18. Ibid., 52.
19. Ibid., 40–41. Interestingly, see Walter Lippmann, *Public Opinion* (New York: Harcourt, Brace and Co., 1922), 173: "The doctrine of the omnicompetent citizen is for most practical purposes true in the rural township."
20. Ibid., 218, 291, 211 (glosses from same pages).

21. Hannah Arendt, *The Human Condition* (Chicago: University of Chicago Press, 1958).
22. Dewey, *Public*, 211, 216–17, 211.
23. Cited in Robert Westbrook, *John Dewey and American Democracy* (Ithaca: Cornell University Press, 1990), 225; also see 457.
24. John Dewey, "Freedom and Culture," in *The Later Works: 1925–1953: Vol. 13, 1938–1939*, ed. Jo Ann Boydston (Carbondale: Southern Illinois University Press, 1988), 176.
25. See Anne Phillips, "Dealing with Difference: A Politics of Ideas or a Politics of Presence?" in *Democracy and Difference: Contesting the Boundaries of the Political*, ed. Seyla Benhabib (Princeton, NJ: Princeton University Press, 1996), 139–52.
26. Mansbridge, "Paradox of Size," 168, 163.
27. Jane J. Mansbridge, *Why We Lost the ERA* (Chicago: University of Chicago Press, 1986).
28. Mansbridge, "Paradox of Size," 172. Dahl and Tufte, *Size and Democracy*, 138, also concluded that "democratic goals conflict and no single unit or kind of unit can best serve these goals."
29. Ibid. Another option Dewey also referred to both in *Public* and in "Freedom and Culture" was that "groups having a functional basis will probably have to replace those based on physical contiguity" (177). It is not clear, however, how this provides a solution any different than that represented by the federation approach.
30. Dewey, *Public*, 185.
31. Alan Ryan, *John Dewey and the High Tide of American Liberalism* (New York: W. W. Norton, 1995), 414.
32. Westbrook, *Dewey*, 316.
33. Katherine Camp Mayhew and Anna Camp Edwards, *The Dewey School: The Laboratory School of the University of Chicago, 1896–1903* (New York: D. Appleton-Century, 1936), 489.
34. John Dewey, *Democracy and Education: An Introduction to the Philosophy of Education* (New York: Macmillan, 1916), 197.
35. Westbrook, *Dewey*, 318.
36. Dewey, "A *Common Faith*," in *The Later Works*, vol. 9, ed. Jo Anne Boydston (Carbondale: Southern Illinois University Press, 1981), 57, 19, 17, italics added.
37. Ibid., 33.
38. Steven C. Rockefeller, *John Dewey: Religious Faith and Democratic Humanism* (New York: Columbia University Press, 1991), 64.
39. Dewey, *Public*, 185.
40. Gail Kennedy, "Pragmatism, Pragmaticism, and the Will to Believe—A Reconsideration," *The Journal of Philosophy* 55, no. 14 (1958): 578–88.
41. Maxine Greene, "Exclusions and Awakenings," in *Learning from Our Lives*, ed. Anna Neumann and Penelope L. Peterson (New York: Teachers College Press, 1997), 18–36; Raymond D. Boisvert, "The Nemesis of Necessity: Tragedy's

Challenge to Deweyan Pragmatism," in *Dewey Reconfigured: Essays of Deweyan Pragmatism*, ed. Casey Haskins and David I. Seiple (Albany: SUNY Press, 1999), 151–68. On the importance of a tragic view of life in pragmatic thought, also see Cornel West, *The American Evasion of Philosophy: A Genealogy of Pragmatism* (Madison: University of Wisconsin Press, 1989).

42. Hannah Arendt, *The Origins of Totalitarianism* (New York: Harcourt, Brace and World, 1966); see Aaron Schutz, "Contesting Utopianism: Hannah Arendt and the Tensions of Democratic Education," in *Preserving Our Common World: Essays on Hannah Arendt and Education*, ed. Mordechai Gordon (Boulder: Westview, 2001), 93–126.

Chapter 4

1. Susan F. Semel and Alan R. Sadovnik, "*Schools of Tomorrow*," *Schools of Today: What Happened to Progressive Education* (New York: Peter Lang, 1999); John Dewey, *Schools of To-Morrow*, ed. Evelyn Dewey (New York: E. P. Dutton, 1915).
2. Robert B. Westbrook, *John Dewey and American Democracy* (Ithaca, NY: Cornell University Press, 1991), 501.
3. Jan Olof Bengtsson, *The Worldview of Personalism* (New York: Oxford University Press, 2006); Rufus Burrow Jr., *Personalism: A Critical Introduction* (St. Louis, MO: Chalice Press, 1999).
4. Bernard Schmidt, "Bronson Alcott's Developing Personalism and the Argument with Emerson," *American Transcendental Quarterly* 8, no. 14 (1994): 311–27.
5. James J. Farrell, *The Spirit of the Sixties: The Making of Postwar Radicalism* (New York: Routledge, 1997) gave an overview of the broad uses of personalism in America from Day's conversion in the 1930s, through the emergence of the Civil Rights Movement and antinuclear pacifism during the 1950s, to the dissipation of the counterculture in the North in the 1970s.
6. With respect to education, see especially Paul Averich, *The Modern School Movement* (Oakland, CA: AK Press, 2006). Numerous personalists considered themselves anarchists at different times, including Waldo Frank.
7. Burrow, *Personalism*, 89, noted that "personalism conceives of reality as through and through social, relational, or communal." In fact, it is possible, Burrow noted, to conceive of "person" in personalism as "the capacity for fellowship," although, "this does not mean . . . that personalism affirms the fundamentality of relationality, as if it has some meaning apart from person" (106).
8. Paul Lichterman, *The Search for Political Community: American Activists Reinventing Commitment* (New York: Cambridge University Press, 1996), 6.
9. This was Harold Rugg's solution to the problem of democratic education within a personalist framework. Ronald W. Evans, *This Happened in America: Harold Rugg and the Censure of Social Studies* (Charlotte, NC: Information Age Pub., 2007).

10. Michael Löwy and Robert Sayre, *Romanticism Against the Tide of Modernity* (Durham, NC: Duke University Press, 2001), 25.
11. Terry P. Pinkard, *German Philosophy, 1760–1860: The Legacy of Idealism* (Cambridge: Cambridge University Press, 2002), 167–68, 169.
12. Philip F. Gura, *American Transcendentalism: A History* (New York: Hill and Wang, 2007); Lance Newman, *Our Common Dwelling: Henry Thoreau, Transcendentalism, and the Class Politics of Nature* (New York: Palgrave Macmillan, 2005).
13. Newman, *Our Common Dwelling*, 34.
14. "What distinguishes the two groups is political education. Brownson and the other reformers had concrete experience of the class fracture in society and of the hard fight to close it. . . . [Reformers spoke from] the crucible of deeply challenging lessons in working-class aspirations and habits of thought. The scholars, on the other hand, were rarely pushed beyond the brahminical pale." Ibid., 110.
15. Bronson Alcott, "Bronson Alcott on Amusements," *Concord Magazine*, March 1999, http://www.concordma.com/magazine/mar99/amuse.html; see Elizabeth Palmer Peabody and Amos Bronson Alcott, *Record of Mr. Alcott's School Exemplifying the Principles and Methods of Moral Culture* (Boston: Roberts, 1888).
16. William E. Leuchtenburg, *The Perils of Prosperity, 1914–1932* (Chicago: University of Chicago Press, 1993), 124.
17. Ibid., 187–88, 198.
18. Lynn Dumenil, *The Modern Temper: American Culture and Society in the 1920s* (New York: Hill and Wang, 1995), 8.
19. Ibid., 57.
20. Ibid., 146.
21. Paula S. Fass, *The Damned and the Beautiful: American Youth in the 1920s* (New York: Oxford University Press, 1979), 19.
22. Ibid., 328, 329, 360.
23. See Robert Cohen, *When the Old Left was Young* (New York: Oxford University Press, 1993).
24. Most of this section is informed by Casey Blake, *Beloved Community: The Cultural Criticism of Randolph Bourne, Van Wyk Brooks, Waldo Frank, and Lewis Mumford* (Chapel Hill: University of North Carolina Press, 1990), which is the most comprehensive text about the Young Americans and other associates like Rosenfeld. See also James Livingston, *Accumulating America: Pragmatism, Consumer Capitalism, and Cultural Revolution, 1850–1940* (Chapel Hill: University of North Carolina Press, 1994). A number of biographies have also informed my analysis, including Bruce Clayton, *Forgotten Prophet: The Life of Randolph Bourne* (Baton Rouge: Louisiana State University Press, 1984); Raymond Nelson, *Van Wyck Brooks: A Writer's Life* (New York: Dutton, 1981); Donald L. Miller, *Lewis Mumford, A Life* (New York: Weidenfeld and Nicolson, 1989); and although it is quite limited, Michael A. Ogorzaly, *Waldo Frank: Prophet of Hispanic Regeneration* (Lewisburg, PA: Bucknell University Press, 1994). Key works

that defined the cultural, social, and political vision of this group included Randolph Bourne, "Trans-National America," in *History of a Literary Radical and Other Essays*, ed. Randolph Bourne (New York: B. W. Huebsch, 1920), 266–99; Van Wyck Brooks, *America's Coming-of-Age* (New York: Viking, 1930); Waldo Frank, *Our America* (New York: Boni and Liveright, 1919); and Lewis Mumford, *The Golden Day: A Study in American Experience and Culture* (New York: Boni and Liveright, 1926). Also see, Christine Stansell, *American Moderns: Bohemian New York and the Creation of a New Century* (New York: Metropolitan Books, 2000), for a discussion of the intellectual vibrancy and radical lifestyles of the Greenwich Village context of the time, and Celeste Connor, *Democratic Visions: Art and Theory of the Stieglitz Circle, 1924–1934* (Berkeley: University of California Press, 2001), for a discussion of Steiglitz and, to a lesser extent, Rosenfeld.

25. Dumenil, *Modern Temper*, 75.
26. Blake, *Beloved Community*, 8, 124, 51.
27. Ibid., 52, 172, 51.
28. Ibid., 94, 174.
29. Ibid., 158, first italics added.
30. Frank, cited in Blake, *Beloved Community*, 264.
31. Thomas Carlyle Dalton, *Becoming John Dewey: Dilemmas of a Philosopher and Naturalist* (Bloomington: Indiana University Press, 2002), 112, 114, 118.
32. John Dewey, "Letter to Max Otto, September 10, 1940," in *The Correspondence of John Dewey (CD-ROM)*, ed. Larry A. Hickman (Charlottesville, VA: Intelex, 1992).
33. Lewis Mumford, *The Golden Day* (New York: Boni and Liveright, 1926).
34. Lewis Mumford, "The Pragmatic Acquiescence," in *Pragmatism and American Culture*, ed. Gail Kennedy (Boston: Heath, 1950), 47. Note that this anthology collects the key texts of this debate as well as other related texts together in an accessible way.
35. Mumford, "The Pragmatic Acquiescence," 48. In my description of this exchange, I have emphasized those aspects of Mumford's argument that seem most compelling. Even so, one can see areas in my summary where he was clearly being unfair to the richness of Dewey's thought. See Westbrook, *Dewey*.
36. John Dewey, "The Pragmatic Acquiescence," in *Pragmatism and American Culture*, ed. Gail Kennedy (Boston: Heath, 1950), 53, 50.
37. Lewis Mumford, "The Pragmatic Acquiescence: A Reply," in *Pragmatism and American Culture*, ed. Gail Kennedy (Boston: Heath, 1950), 54.
38. Ibid., 54–55.
39. Ibid., 56.
40. Ibid., 56, 57, 56.
41. Waldo David Frank, *The Re-discovery of America: An Introduction to a Philosophy of American Life* (Westport, Conn.: Greenwood Press, 1929).
42. John Dewey, "Individualism Old and New," in *The Later Works 1925–1953: 1929–1930, vol. 5*, ed. Jo Ann Boydston (Carbondale: Southern Illinois University Press, 1984), 74, 72.

43. Despite his growing commitment to the antiwar movement in the years that followed, Dewey never publicly or, as far as we know, privately apologized (see Westbrook, *Dewey*).
44. Harold Ordway Rugg and Ann Shumaker, *The Child-Centered School: An Appraisal of the New Education* (Yonkers-on-Hudson, NY: World Book Company, 1928); Stanwood Cobb, *The New Leaven: Progressive Education and Its Effect Upon the Child and Society* (New York: John Day, 1928). Key works on the educational vision of the personalists include Robert Beck, "Progressive Education and American Progressivism: Margaret Naumburg," *Teachers College Record* 60, no. 4 (1959): 198–208; Robert Beck, "Progressive Education and American Progressivism: Caroline Pratt," *Teachers College Record* 60, no. 3 (1958): 129–37; Lawrence A. Cremin, *The Transformation of the School: Progressivism in American Education, 1876–1957* (New York: Knopf, 1961).
45. Also see Cobb, *The New Leaven*, and Agnes De Lima, *Our Enemy, The Child* (New York: Arno, 1925).
46. See Stansell, *American Moderns*.
47. Dalton, *Becoming John Dewey*.
48. Beck, "American Progressivism: Naumburg."
49. See the recent biography of Pratt by Mary E. Hauser, *Learning from Children: The Life and Legacy of Caroline Pratt* (New York: Peter Lang, 2006). This work provides a useful overview of Pratt's life, but lacks much reference to the context of her life in New York.
50. D. B. Curtis, "Psychoanalysis and Progressive Education: Margaret Naumburg at the Walden School," *Vitae Scholasticae* 2, no. 2 (1983): 354.
51. Beck, "American Progressivism: Naumburg," 204.
52. Miss Levin, from Pratt's City and Country School, cited in Rugg and Shumaker, *Child-Centered School*, 229.
53. Naumburg, *The Child and the World: Dialogues in Modern Education* (New York: Harcourt Brace, 1928), 289–90.
54. Ibid., 306–7.
55. Ibid., 121.
56. Ibid., 121–22.
57. Cited in Curtis, "Psychoanalysis and Progressive Education," 354.
58. Rugg and Shumaker, *Child-Centered School*, 65.
59. Hauser, *Life and Legacy of Caroline Pratt*, 77.
60. John Dewey, "Progressive Education and the Science of Education," in *The Later Works: 1925–1953: Vol. 3, 1927–1928*, ed. Jo Ann Boydston (Carbondale: Southern Illinois University Press, 1984), 262–63, 266.
61. Beck, "American Progressive Education: Naumburg."
62. Beck, "American Progressive Education: Naumburg," 204.
63. Naumburg, *Child and the World*, 58, italics added.
64. Ibid., 59–60, italics added.
65. Ibid., 59, italics added.
66. Ibid., 111.

67. In 1922, he acknowledged that in "trying to stand for freedom, I have found that I have been considered by many as upholding the doctrine 'that children should do exactly as they please.' Therefore I was led to analyze more carefully what I really did believe, and I had to admit that I had set out without knowing what the real meaning of individuality was." John Dewey, "What I Believe, Revised," in *The Middle Works 1899–1924: Volume 15, 1923–24*, ed. Jo Ann Boydston (Carbondale: Southern Illinois University Press, 1983), 178.
68. While he noted at one point that he was talking about the most extreme examples of personalist pedagogy, most of his statements clearly referred to the movement more broadly.
69. Dewey, John Dewey, "Individuality and Experience," in *Art and Education*, ed. John Dewey and Albert C. Barnes (Merion, PA: The Barnes Foundation Press, 1929), 37–38.
70. Ibid., 33.
71. Ibid., 32–33.
72. John Dewey, "How Much Freedom in the New Schools?" in *The Later Works: 1925–1953: Vol. 5, 1929–1930*, ed. Jo Ann Boydston (Carbondale: Southern Illinois University Press, 1984), 323.
73. Ibid., 323–4, 323, 325,
74. Dalton, *Becoming John Dewey*.
75. See David A. Granger, *John Dewey, Robert Pirsig, and the Art of Living: Revisioning Aesthetic Education* (New York: Palgrave Macmillan, 2006); Naoko Saito, *The Gleam of Light: Moral Perfectionism and Education in Dewey and Emerson* (New York: Fordham University Press, 2005).
76. John Dewey, *Experience and Education* (New York: Macmillan, 1938), 63.
77. Hauser, *Life and Legacy of Caroline Pratt*, 77.
78. Rugg and Shumaker, *Child-Centered School*, 115.
79. See n. 45.
80. John Dewey, "Individuality in Education," in *The Middle Works: 1899–1925: Vol. 15, 1923–1924*, ed. Jo Ann Boydston (Carbondale: Southern Illinois University Press, 1983), 176.
81. Personalists like Naumburg also often delayed reading, but for different reasons.
82. Dewey, "What I Believe, Revised," 31–35.
83. Lawrence Cremin, *The Transformation of the School*; Beck, "Progressive Education: Naumburg"; Beck, "Progressive Education: Pratt"; Hauser, *Life and Legacy of Caroline Pratt*.

Chapter 5

1. Alexander Sutherland Neill, *Summerhill: A Radical Approach to Child Rearing* (New York: Hart, 1960); Alexander Sutherland Neill, *Freedom—Not License!* (New York: Hart, 1966).
2. James J. Farrell, *The Spirit of the Sixties: Making Postwar Radicalism* (New York: Routledge, 1997), 205.

3. Ibid., 229.
4. Burton Weltman, "Revisiting Paul Goodman: Anarcho-Syndicalism as the American Way of Life," *Educational Theory* 50, no. 2 (2000): 181.
5. Another exception was George Dennison, whose *The Lives of Children* (Reading, MA: Addison-Wesley, 1969) was perhaps the most lyrical defense of free schools in America and was also somewhat unique in its focus on poor children. Dennison and Goodman were closely connected, since Goodman was for a time Dennison's therapist, and Goodman's daughter worked at Dennison's school.
6. Taylor Stoehr, *Here Now Next: Paul Goodman and the Origins of Gestalt Therapy* (New York: The Analytic Press, 1997), 76.
7. Neill, *Summerhill*, 4.
8. Alexander Sutherland Neill, *The Free Child* (London: Jenkins, 1953), 29.
9. John Dewey, "Human Nature and Conduct," in *The Middle Works: 1899–1925: Vol. 14, 1922*, ed. Jo Ann Boydston (Carbondale: Southern Illinois University Press, 1983), 69. Subsequently cited as *HNC*.
10. Ibid., 65.
11. Ibid., 109.
12. Ibid., 113.
13. See, for example, his discussion in John Dewey, *The Child and the Curriculum; and The School and Society* (Chicago: University of Chicago Press, 1956).
14. Kris Gutierrez, Betsy Rymes, and Joanne Larson, "Script, Counterscript, and Underlife in the Classroom," *Harvard Educational Review* 65, no. 3 (1995): 445–71.
15. John Dewey, "The Reflex Arc Concept in Psychology," *The Early Works: 1882–1898: Vol. 1, 1882–1888*, ed. Jo Ann Boydston (Carbondale: Southern Illinois University Press, 1972), 96–110.
16. Dewey, *HNC*, 89, 71–72.
17. Frederick S. Perls, Ralph E. Hefferline, and Paul Goodman, *Gestalt Therapy: Excitement and Growth in the Human Personality* (New York: Dell, 1951). While Goodman was not the first author of *Gestalt Therapy*, evidence indicates that he was the one who largely wrote the theoretical chapters, informed by Perls's ideas. Hefferline wrote the early chapters discussing exercises he had his students at Columbia University engage in. For more information about the authorship of *Gestalt Therapy*, see Stoehr, *Origins of Gestalt Therapy*.
18. Stoehr, *Origins of Gestalt Therapy*, 266.
19. Ron Miller, *Free Schools, Free People: Education and Democracy After the 1960s* (Albany: State University of New York Press, 2002).
20. For example, see Goodman, *Gestalt Therapy*, 215.
21. See, for example, John Dewey, *Democracy and Education: An Introduction to the Philosophy of Education* (New York: Macmillan, 1916).
22. Goodman, *Growing Up Absurd: Problems of Youth in the Organized System* (New York: Random House, 1960), 73.
23. Goodman, *Compulsory Mis-Education and the Community of Scholars* (New York: Vintage, 1964), 23, 259.

24. Goodman, *Growing Up Absurd*, 145.
25. Ibid., 88, 73.
26. Goodman, *Compulsory Mis-Education*, 23.
27. Ibid., 67, 45.
28. Goodman, *Gestalt Therapy*, 152.
29. Ibid., x.
30. See Michel Foucault, *Discipline and Punish: The Birth of the Prison* (New York: Pantheon Books, 1977).
31. Goodman, *Compulsory Mis-Education*, 8.
32. Other radical intellectuals of the time were not as reticent as Goodman about emphasizing the comparative freedom and, in some cases, the cultural superiority of those who were most oppressed in modern America. Friedenberg argued, for example, that "if effective," the civil rights movement "will squander the last aristocratic element that might have countervailed, however feebly, against the severe and debilitating demosis that now cripples our culture." He appeared to recommend that African American leaders cling to their oppression so some small number of them might continue to "remind us" (e.g., successful members of the middle class) "of what dignity is." Edgar Friedenberg, *Coming of Age in America* (New York: Random House, 1965), 243, 242.
33. Nathan Glazer and Paul Goodman, "Berkeley: An Exchange," *New York Review of Books* 4, no. 1 (1965): 22.
34. See Aaron Schutz, "Rethinking Domination and Resistance: Challenging Postmodernism," *Educational Researcher* 76, no. 4 (2004): 691–743.
35. Theodore Roszak, *The Making of a Counter Culture: Reflections on the Technocratic Society and Its Youthful Opposition* (Berkeley: University of California Press, 1995).
36. Jonathan Croall, *Neill of Summerhill: The Permanent Rebel* (London: Routledge and K. Paul, 1983), 436.
37. The exception to this was his early work with Homer Lane who developed an effective "unconditional regard" strategy when working with juvenile delinquents. But whether this would lead to success in the world outside, by itself, is doubtful. The work of Goodman's former psychology client and friend George Dennison, described in *The Lives of Children*, where the staff actually did seek to actively educate, albeit in a "free school" manner, seems like a more relevant comparison.
38. Neill, *Summerhill*, 33, italics added. Also see 116.
39. Pierre Bourdieu, *Practical Reason: On the Theory of Action* (Cambridge: Polity, 1998).
40. Ray Hemmings, *Fifty Years of Freedom: A Study of the Development of the Ideas of A. S. Neill* (London: Allen and Unwin, 1972), 133.
41. Goodman, *Gestalt Therapy*, 376, 246.
42. Ibid., 35, 259.
43. Cited in Weltman, "Revisiting Paul Goodman," 6.
44. Goodman, *Gestalt Therapy*, 379.
45. John Dewey, *Democracy and Education* (New York: Macmillan, 1916), 14.

46. Ibid., 18, italics added, 51.
47. One reason why Dewey may have thought the emergence of natural cooperation was unlikely was his belief that humans are born with "an original tendency to assimilate objects and events to the self, to make them part of the 'me.'" In fact, "we may even admit that the 'me' cannot exist with out the 'mine.' The self gets solidity and form through an appropriation of things which identifies them with whatever we call myself." While this tendency seemed undeniable, as with impulse energy, he noted that it could express itself in a vast range of ways. "The institution of private property," he argued, "is not the only way to appropriate" aspects of the world to the self. One could, for example, change "mine" to "my community," which, of course, was exactly what he wanted to do with democracy. Dewey, *HNC*, 82–83.
48. Dewey, *Democracy and Education*, 108.
49. Ibid., 133.
50. John Dewey, *Experience and Education* (New York: Macmillan, 1938), 56, 60.
51. Stoehr, *Origins of Gestalt Therapy*, 16.
52. Goodman, *Gestalt Therapy*, 309, 247, 309.
53. Hemmings, *Fifty Years of Freedom*, 41.
54. Neill, *Summerhill*, 4, 106, 250–51. "My own opinion, after thirty-one years of Summerhill, is that freedom alone will cure most delinquencies in a child. Freedom of course, not license, not sentimentality, not idealism. Freedom alone will not cure pathological cases; it will barely touch cases of arrested development" (88).
55. "If baby plays with the cat and gets scratched it is not punishment; if it plays with matches and gets burned it is not being punished. Natural reactions do not constitute punishment, because there is no suggestion of right or wrong in them: punishment implies the judgment of an authority." Neill, *Free Child*, 63.
56. Neill, *Summerhill*, 8, 144. Note the fundamentally "ownership society" vision of this model. Property is assumed to belong to someone specific all through *Summerhill*, in contrast with many other "free" educational contexts. This would be a fascinating issue for someone to pursue.
57. Ibid., 25, 26, 29.
58. Ibid., 46.
59. Ibid., 53, 114.
60. Neill cited in Hemmings, *Fifty Years of Freedom*, 72.
61. Goodman, *Gestalt Therapy*, 309.
62. Theodore Roszack, *The Making of a Counter Culture: Reflections on the Technocratic Society and Its Youthful Opposition* (New York: Anchor Books, 1969), 198.
63. Goodman, *Growing Up Absurd*, 107.
64. Stoehr, *Origins of Gestalt Therapy*, 19.
65. Taylor Stoehr, *Decentralizing Power: Paul Goodman's Social Criticism* (Montreal: Black Rose Books, 1994), xiv.
66. Miller, *Free Schools*, 33, 160, 161, 75.

67. The ERAP effort in Cleveland was one exception, and SDS's work there helped provide the foundation for one of the strongest local National Welfare Rights Association community organizing groups a few years later.
68. James Miller, *Democracy Is in the Streets* (Cambridge, MA: Harvard University Press, 1994).
69. Jonathan Kozol, *Free Schools* (Boston: Houghton Mifflin, 1972), 12.
70. See Lisa D. Delpit, *Other People's Children: Cultural Conflict in the Classroom* (New York: New Press, 1995).
71. Kozol, *Free Schools*, 60.
72. Ibid., 93–94.
73. Also see Barbara Leslie Epstein, *Political Protest and Cultural Revolution: Nonviolent Direct Action in the 1970s and 1980s* (Berkeley: University of California Press, 1991).
74. Paul Lichterman, *The Search for Political Community: American Activists Reinventing Commitment* (New York: Cambridge University Press, 1996), 46, 48, 194. See Epstein, *Political Protest and Cultural Revolution* for a similar discussion.
75. Epstein, *Political Protest and Cultural Revolution*.
76. Roszak, *The Making of a Counter Culture*, 200.
77. Weltman, "Revisiting Paul Goodman," 179–99.
78. David F. Labaree, *The Trouble with Ed Schools* (New Haven, CT: Yale University Press, 2004).
79. Aaron Schutz, "Caring in Schools is not Enough: Community, Narrative, and the Limits of Alterity," *Educational Theory* 48, no. 3 (1998): 373–93.

Chapter 6

1. It is important to note that models of community organizing are significantly different from those informing unions. As Alinsky noted, "Labor union organizers turned out to be poor community organizers. Their experience was tied to a pattern of fixed points, whether it was definite demands on wages, pensions, vacation periods, or other working conditions, and all of this was anchored into particular contract dates." Saul Alinsky, *Rules for Radicals* (New York: Vintage, 1971), 66.
2. Donald C. Reitzes and Dietrich C. Reitzes, *The Alinsky Legacy: Alive and Kicking* (Greenwich, CT: Jai Press, 1987); Heidi J. Swarts, *Organizing Urban America: Secular and Faith-Based Progressive Movements* (Minneapolis: University of Minnesota Press, 2008); Edward T. D. Chambers and Michael A. Cowan, *Roots for Radicals: Organizing for Power, Action, and Justice* (New York: Continuum, 2003).
3. Aaron Schutz, "Core Dilemmas of Community Organizing" series at the Open Left blog, collected at http://www.educationaction.org/core-dilemmas-of-community-organizing.html.
4. Sandford Horwitt, *Let Them Call Me Rebel: Saul Alinsky: His Life and Legacy* (New York: Vintage, 1992). Alinsky often told interviewers that he had

grown up in a "Jewish slum." Horwitt, however, found that "living in a poor, overcrowded ghetto was, at the most, an experience confined to Saul's earliest childhood" (6).
5. Fred Rose, *Coalitions Across the Class Divide: Lessons from the Labor, Peace, and Environmental Movements* (Ithaca, NY: Cornell University Press, 2000).
6. Reitzes and Reitzes, *The Alinsky Legacy*, 5.
7. Saul Alinsky, *Reveille for Radicals* (New York: Vintage, 1946), 134, 133, 134.
8. Ibid., ix.
9. Ibid., 133, 132, 135.
10. Alinsky, *Rules*, 21.
11. Ibid., 116.
12. Ibid., 134.
13. Alinsky, *Reveille*, 229.
14. Ibid., xiv.
15. See Horwitt, *Let Them Call Me Rebel*.
16. Alinsky, *Rules*, 74.
17. Von Hoffman cited in Horwitt, *Let Them Call Me Rebel*, 397.
18. Alinksy, *Reveille*, 54–55.
19. Ibid., 64–65, italics added to second.
20. Ibid., 158, original italics removed from entire first quote.
21. Paul Lichterman, *The Search for Political Community: American Activists Reinventing Commitment* (New York: Cambridge University Press, 1996), 120–21, 123.
22. The following discussion draws from my own experience in a congregational organizing group, as well as from a range of key works on organizing. Key academic case studies include Swarts, *Organizing Urban America*; Mark R. Warren, *Dry Bones Rattling: Community Building to Revitalize American Democracy* (Princeton, NJ: Princeton University Press, 2001); Dennis Shirley, *Community Organizing for Urban School Reform* (Austin: University of Texas Press, 1997). Examples of works on organizing practice include Michael Jacoby Brown, *Building Powerful Community Organizations* (Arlington, MA: Long Haul, 2006); Si Kahn, *Organizing* (New York: McGraw-Hill, 1982); Kimberley A. Bobo et al., *Organizing for Social Change*, 3rd ed. (Santa Ana, CA: Seven Locks, 2001); Chambers and Cowan, *Roots for Radicals*; Lee Staples, *Roots to Power*, 2nd ed. (Westport, CT: Praeger, 2004).
23. See Bernard Bass and Ronald Riggio, *Transformational Leadership*, 2nd ed. (New York: Lawrence Erlbaum, 2005).
24. Ron Miller, *Free Schools, Free People: Education and Democracy After the 1960s* (Albany: State University of New York Press, 2002).
25. Barbara Leslie Epstein, *Political Protest and Cultural Revolution: Nonviolent Direct Action in the 1970s and 1980s* (Berkeley: University of California Press, 1991); Lichterman, *Political Community*.
26. Benjamin R. Barber, *Strong Democracy* (Berkeley: University of California Press, 2003), 238.

27. Point made by Sirianni, taken from Matthew L. Trachman, *Rethinking Leadership: Presidential Leadership and the "Spirit of the Game" of Democracy* (North York, ON: York University, 1999). See Jo Freeman, "The Tyranny of Structurelessness," http://www.jofreeman.com/joreen/tyranny.htm (accessed February 4, 2010). There are exceptions, of course. See writings on the AIDS activism group ACT-UP, for example, Michael Brown, *Replacing Citizenship: AIDS Activism and Radical Democracy* (New York: Guilford, 1997).
28. From Trachman, *Rethinking Leadership*, 2.
29. Dahl cited in John Kane, "The Ethical Paradox of Democratic Leadership," *Taiwan Journal of Democracy* 3, no. 2 (2007): 34.
30. Kenneth Patrick Ruscio, *The Leadership Dilemma in Modern Democracy* (Northampton, MA: Edward Elgar, 2004), ix.
31. Kane, "Ethical Paradox," 35.
32. See, for example, Ruscio, *Leadership Dilemma*; Trachman, *Rethinking Leadership*; J. Thomas Wren, *Inventing Leadership: The Challenge of Democracy* (Cheltenham, UK: Edward Elgar, 2007).
33. Roderick Bush, *We Are Not What We Seem: Black Nationalism and Class Struggle in the American Century* (New York: New York University Press, 2000), 185.
34. Nicholas von Hoffman, *Finding and Making Leaders* (Chicago: Industrial Areas Foundation, 1963), 11.
35. Alinsky, *Reveille*, 155, 158.
36. Ibid., 158.
37. Ibid., 129.
38. Ibid., 184.
39. Ibid., 159.
40. Ibid., 170.
41. Ibid., 158–59.
42. Alinsky, *Rules*, 106.
43. Joseph Kahne and Joel Westheimer, "The Limits of Political Efficacy: Educating Citizens for a Democratic Society," *PS: Political Science and Politics* 39 (2006): 289–96.
44. Alinsky, *Reveille*, 77.
45. Alinsky, *Rules*, 100.
46. Alinsky, *Reveille*, 122.
47. Alinsky, *Rules*, 112, 71.
48. Alinsky, *Reveille*, 164–65.
49. Ibid., 97–98.
50. See Saul Alinsky, "Empowering People Not Elites: An Interview with Saul Alinsky," *Playboy* (1972). The full interview is available at http://britell.com/alinsky.html, and my citations are taken from that page.
51. Alinsky, *Rules*, xi.
52. Alinsky, *Reveille*, 186.
53. Ibid., 134.
54. Alinsky, *Rules*, 36.
55. Law, *Wolf Shall Dwell with the Lamb*, 65.

56. "Core Dilemmas of Community Organizing" series at the Open Left blog, http://www.educationaction.org/core-dilemmas-of-community-organizing.htm.

Chapter 7

1. Aaron Schutz, "Misunderstanding Mississippi: The Freedom Schools and Personalist Pedagogy," paper presented at the 2004 American Educational Research Association Conference, San Diego, California, http://www.educationaction.org/progressive-democracy-book-extras.html (accessed May 24, 2010); and "Education Scholars Have Much to Learn About Social Action: An Essay Review," *Education Review* 10, no. 3, retrieved May 24, 2010, from the previously mentioned website and also http://edrev.asu.edu/essays/v10n3index.html.
2. Michael Eric Dyson, *I May Not Get There With You: The True Martin Luther King, Jr.* (Free Press, 2001).
3. This is much less evident in writings about the civil rights movement by educational historians like Daniel Perlstein. What I am mostly referring to, here, are more general references to and uses of the civil rights movement in a range of less rigorous historical writings in education.
4. Aaron Schutz, "Creating Local 'Public Spaces' in Schools: Insights from Hannah Arendt and Maxine Greene," *Curriculum Inquiry* 29, no. 1 (1999): 77–98.
5. Terry H. Anderson, *The Movement and the Sixties* (New York: Oxford University Press, 1996).
6. I am including rural farmers in the "working class" in this chapter, as have some other scholars of the movement. While this brings some problems with it, for the purposes of this chapter it suffices.
7. Barbara Ransby, *Ella Baker and the Black Freedom Movement: A Radical Democratic Vision* (Durham: University of North Carolina Press, 2005), 270.
8. Ibid., 7, 142.
9. See Frank Adams, *Unearthing Seeds of Fire: The Idea of Highlander* (New York: John F. Blair, 1975); Myles Horton et al., *We Make the Road by Walking: Conversations on Education and Social Change* (Philadelphia, PA: Temple University Press, 1990); Barbara Thayer-Bacon, "An Exploration of Myles Horton's Democratic Praxis: Highlander Folk School," *Educational Foundations* 18, no. 2 (2004): 5–23.
10. Ransby, *Ella Baker*, 139, 274, 305.
11. Ibid., 365.
12. Ibid., 249–50.
13. Ibid., 368, 309, 369, 188.
14. Francesca Polletta, "Contending Stories: Narrative in Social Movements," *Qualitative Sociology* 21 (1998): 152, 271.
15. Ransby, *Ella Baker*, 188, 191.
16. Eric Burner, *And Gently He Shall Lead Them: Robert Parris Moses and Civil Rights in Mississippi* (New York: New York University Press, 1994), 2, 4, 6–7, 70.

17. Clayborne Carson, *In Struggle: SNCC and the Black Awakening of the 1960s* (Cambridge, MA: Harvard University Press, 1995), 30.
18. Ransby, *Ella Baker*, 130, 261.
19. Francesca Polletta, "'It Was Like a Fever...' Narratives in Social Protest," *Social Problems* 45, no. 2 (1998): 152. See also Heidi J. Swarts, *Organizing Urban America: Secular and Faith-Based Progressive Movements* (Minneapolis: University of Minnesota Press, 2008). This discomfort with institutionalization also played into the disappearance of Baker's rich vision of social struggle from the collective memory of those who came after her. Her model of empowerment was enacted in real relationships, one-to-one. But just as public spaces ungrounded by institutional structures disappear the moment people withdraw, so are visions of organizing, no matter how sophisticated or profound, likely to be lost if they are not codified in some way, however problematically.
20. Ransby, *Ella Baker*, 280. Early in his career, Alinsky argued that organizing groups could only last a few years. Later in his life, his biographer reported, Alinsky realized that he had been wrong about this. See Sandford Horwitt, *Let Them Call Me Rebel: Saul Alinsky: His Life and Legacy* (New York: Vintage, 1992).
21. There has been much discussion in the literature on social movements about whether the move to institutionalization blocks social action. This was most famously argued by Frances Fox Piven and Richard A. Cloward, *Poor People's Movements: Why They Succeed, How They Fail* (New York: Vintage Books, 1979). Later on, however, even they seem to have moderated their position some, arguing that the danger of "organizations" emerged more in specific moments of mass rebellion. In contrast, William A. Gamson, in *The Strategy of Social Protest* (Homewood, IL: Dorsey, 1975), found that organizational structure was correlated with success across the history of social action in America. Work on the civil rights movement, especially Aldon D. Morris, *Origins of the Civil Rights Movements* (New York: Free Press, 1986), stresses how important existing structures like the African American church were to the success of movements, but that this was often obscured in favor of a myth of spontaneous action. In any case, without some structure to maintain the lessons of a movement, they are in danger of being lost. Piven appears to acknowledge this, and Gamson's conclusions in particular, in her most recent book, *Challenging Authority: How Ordinary People Change America* (New York: Rowan and Littlefield, 2008).
22. Ransby, *Ella Baker*, 113, 115.
23. Ibid., 274.
24. Burner, *Robert Moses*, 54.
25. Ransby, *Ella Baker*, 318.
26. Ibid., 217.
27. Taylor Branch, *Parting the Waters: America in the King Years, 1954–63* (New York: Simon and Schuster, 1998).
28. Ibid., 194.
29. Morris, *Origins*, 194.

30. Robert Moses and Charles Cobb Jr., *Radical Equations* (New York: Beacon, 2001), 81.
31. Morris, *Origins*, 194.
32. Richard H. King, *Civil Rights and the Idea of Freedom* (New York: Oxford University Press, 1992), 47.
33. Ibid., 42.
34. Charles M. Payne, *I've Got the Light of Freedom: The Organizing Tradition and the Mississippi Freedom Struggle* (Berkeley: University of California Press, 2007), 260.
35. Referencing William Hinton's work on China in Payne, *Light of Freedom*, 261.
36. Payne, *Light of Freedom*, 238.
37. Carson, *In Struggle*, 143, 202, 143, 171.
38. Ibid., 155
39. Carson, *In Struggle*, 171.
40. Adam Fairclough, *To Redeem the Soul of America: The Southern Christian Leadership Conference and Martin Luther King, Jr.* (Athens: University of Georgia Press, 1987), 18.
41. Gayraud Wilmore, *Black Religion and Black Radicalism*, 3rd ed. (New York: Orbis, 1998), 207. In the end, only a small minority of the religious community responded to this call, most remaining on the sidelines or, in some cases, actively opposing the movement. In Birmingham, for example, "one of the most highly organized and effective of all the civil rights campaigns—only about 20 of the city's 250 black ministers actively supported SCLC" (Fairclough, *To Redeem the Soul of America*, 35).
42. Morris, *Origins*, 9.
43. Gayraud S. Wilmore, *Black Religion and Black Radicalism*, 328. See Fairclough, *To Redeem the Soul of America*.
44. Morris, *Origins*, 77, 7.
45. Fairclough, *To Redeem the Soul of America*, 4.
46. Branch, *Parting the Waters*, 229.
47. Morris, *Origins*, 46.
48. Branch, *Parting the Waters*, 164.
49. Lance Hill, *The Deacons for Defense: Armed Resistance and the Civil Rights Movement* (Durham: University of North Carolina Press, 2006), 15.
50. Interpretation of Bennett in Morris, *Origins*, 47.
51. King, *Idea of Freedom*, 42. Also see, Polletta, "It Was Like a Fever."
52. Fairclough, *To Redeem the Soul of America*, 35.
53. See Mary Patillo's discussion of the position of the middle class in the African American community in Mary Patillo, *Black on the Block: The Politics of Race and Class in the City* (Chicago: University of Chicago Press, 2008).
54. Morris, *Origins*, 8.
55. Ibid., 10.
56. Roderick Bush, *We Are Not What We Seem: Black Nationalism and Class Struggle in the American Century* (New York: New York University Press, 2000), 185.

57. Morris, *Origins*, 10–11. Despite SNCC's complaints, then, King's view of the Movement usually enabled him to rise above narrow organizational interests. As Robinson reported, "he was not an empire builder.... He was interested in the movement and the organization was just a means of pushing the movement. He always struck me as having extraordinary little ego," Morris, *Origins*, 136.
58. Morris, *Origins*, 96, insert from same page.
59. Ibid., 158.
60. Ibid., 158
61. William W. Sales, *From Civil Rights to Black Liberation: Malcom X and the Organization of Afro-American Unity* (New York: Pathfinder, 1994), 47.
62. Sudarshan Kapur, *Raising Up a Prophet: The African-American Encounter with Gandhi* (Boston, MA: Beacon, 1992), 122.
63. Ibid., 123.
64. Morris, *Origins*, 129, 131, 130.
65. Branch, *Parting the Waters*, 179–80.
66. Steve Estes, *I Am a Man! Race, Manhood, and the Civil Rights Movement* (Chapel Hill: University of North Carolina Press, 2005), 79.
67. Hill, *Deacons*, 260–61.
68. Ibid., 27–28, 266, 28.
69. Ibid., 2.
70. Estes, *I Am a Man*, 79.
71. Branch, *Pillar of Fire: America in the King Years, 1963–65* (New York: Simon and Schuster, 1999), 459.
72. Hill, *Deacons*, 259.
73. Ibid. As in all other areas of the civil rights movement, local people did not simply accept the instructions they were given by nonviolence experts as the "truth," instead, they often appropriated this idea to fit the realities of their lives and the understandings that they brought with them.
74. Morris, *Origins*, 158.
75. Branch, *Pillar of Fire*, 195.
76. Dyson, *I May Not Get There*, 85.
77. Estes, *I Am a Man*, 78.
78. See Jeffrey O. G. Ogbar, *Black Power: Radical Politics and African American Identity* (Baltimore, MD: Johns Hopkins University Press, 2005).
79. Hill, *Deacons*, 4–5.
80. Ibid., 16.
81. Ibid., 3–4.
82. Ibid., 268.
83. Dyson, *I May Not Get There*, 86.
84. Hill, *Deacons*, 50.
85. Ibid., 264–65.

Chapter 8

1. See Aaron Schutz, "Core Dilemmas of Community Organizing," Open Left, http://www.educationaction.org/core-dilemmas-of-community-organizing.html (accessed May 28, 2010).
2. Aaron Schutz, "Home Is a Prison in the Global City: The Tragic Failure of School-Based Community-Engagement Strategies," *Review of Educational Research* 76, no. 4 (2006): 699. Quotation from Pedro Noguera, *Racial Isolation, Poverty, and the Limits of Local Control as a Means for Holding Public Schools Accountable* (Los Angeles: UCLA, Institute for Democracy, Education, & Access, 2002), 14, 341–50, http://repositories.cdlib.org/idea/wws/wws-rr011-1002 (accessed May 29, 2010).
3. See Cornel West, *Race Matters* (New York: Vintage, 1994).
4. Listen, Inc., *From the Frontlines: Youth Organizers Speak* (Los Angeles, CA: Privately printed, 2003).
5. A team of graduate students and I examined some of these questions in a local inner-city charter high school, exploring strategies for engaging students in social action. A draft of a paper discussing this experience can be found at http://www.educationaction.org/progressive-democracy-book-extras.html.
6. Todd DeStigter, "Public Displays of Affection: Political Community through Critical Empathy," *Research in the Teaching of English* 33, no. 3 (1999): 235–44; David Stovall, "Forging Community in Race and Class: Critical Race Theory and the Quest for Social Justice in Education," *Race, Ethnicity, and Education* 9, no. 3 (2006): 243–59; Rahima C. Wade, "Service-Learning for Social Justice in the Elementary Classroom: Can We Get There from Here?" *Equity and Excellence in Education* 40, no. 2 (2007): 156–65; Brian D. Schultz, *Spectacular Things Happen Along the Way: Lessons from an Urban Classroom* (New York: Teachers College Press, 2008).
7. Anne Showstack Sassoon, *Gramsci's Politics* (Minneapolis: University of Minnesota Press, 1987), 124.
8. Ibid., 141; Antonio Gramsci, *Prison Notebooks*, trans. Joseph A. Buttigieg (New York: Columbia University Press, 1992).
9. Patricia Hill Collins, *Black Feminist Thought: Knowledge, Consciousness, and the Politics of Empowerment* (New York: Allen and Unwin, 1990); C. L. Barney Dews and Carolyn Leste Law, *This Fine Place So Far from Home: Voices of Academics from the Working Class* (Philadelphia, PA: Temple University Press, 1995); Michelle M. Tokarczyk and Elizabeth A. Fay, *Working-Class Women in the Academy: Laborers in the Knowledge Factory* (Amherst: University of Massachusetts Press, 1993); Alfred Lubrano, *Limbo: Blue-Collar Roots, White-Collar Dreams* (Hoboken, NJ: Wiley, 2004).
10. Annette Lareau, *Unequal Childhoods: Class, Race, and Family Life* (Berkeley: University of California Press, 2003); Aaron Schutz, "Home Is a Prison in the Global City."
11. Because, as a theory professor, I could achieve tenure with research drawn from books, I was able to make the choice to not do formal research as part

of my community organizing work, such as it is. While I do write about organizing, I make no formal effort to collect "data." This has allowed me to not constantly look over my own shoulder, so to speak. I can to be my limited self as a "leader." I am no organic intellectual, not even that great a bridge builder, frankly. And I am not an organizer by any stretch of the imagination. But I think leaving the research behind has allowed me to see organizing from the "inside" better, at least for myself. I think I will mostly keep it that way.

12. Fred Rose, *Coalitions across the Class Divide: Lessons from the Labor, Peace, and Environmental Movements* (Ithaca, NY: Cornell University Press, 2000), 162.
13. Ibid., 176, 178.
14. Paul Lichterman, *The Search for Political Community: American Activists Reinventing Commitment* (New York: Cambridge University Press, 1996).

Index

Addams, Jane, 32, 52–53, 68, 164, 237, 241
administrative progressivism, 3–4, 10, 28–32, 36, 45, 62, 72, 74, 91, 93, 174
aesthetics, 35, 62, 64, 100, 103, 105–6, 117–18, 122, 238, 249
African Americans, 3, 65–67, 94, 141, 153, 169, 178, 183–84, 193–94, 200, 202–17, 251, 257–59
Albany, Georgia, 200, 211–12, 232, 238, 240, 245, 250, 255
Alcott, Bronson, 93, 96, 245, 246
 and School for Human Culture, 96
Alinsky, Saul, 5, 10, 27, 37–39, 67, 153, 161–83, 186, 192, 196, 202, 206–7, 221–26, 238, 253–57
 and Back of the Yards Neighborhood Council, 81, 204
 on David and Roger story, 180
 and ethnography, 163, 165
 and Industrial Areas Foundation (IAF), 172, 174
 on liberals, 37, 153, 163–64, 181–83
 on means and ends, 181–83
 on native leaders, 37, 168, 170, 202
 and *Reveille for Radicals*, 161–63, 167, 238, 254–56
 and *Rules for Radicals*, 183, 221, 253–56
 on self-interest, 42, 179, 180
anarchism, 93, 100, 129–30, 132, 151, 201, 245
Anderson, Terry H., 256
antinuclear activism, 154, 173, 245
Anyon, Jean, 230–32, 234, 236

Apple, Michael, 47, 231–32, 240
Arendt, Hannah, 80, 88, 190, 229, 244–45, 256
Aristotle, 71
art, 64, 101, 103–8, 112–16, 122
Averich, Paul, 245

Babbitt, 98–100, 103
Back of the Yards, 163, 178
Baker, Ella, 192–201, 204, 207, 212, 256, 257–58
Barber, Benjamin R., 173, 255
Bass, Bernard, 234, 254
Bauman, Zygmunt, 23, 235
Beck, Robert, 107–8, 124, 248, 249
Beck, Ulrich, 74, 243
Bengtsson, Jan Olof, 245
Benson, Susan Porter, 25, 232–33, 235
Berkman, Alexander, 100
Bernstein, Basil B., 24, 234–36
Birmingham, Alabama, 197–98, 200, 211–12, 258
Blake, Casey N., 100–101, 238, 246–47
Blumin, Stuart M., 13, 15, 231–33
Bobbio, Norberto, 71, 243
Bobo, Kimberley A., 254
Boisvert, Raymond D., 245
Bourdieu, Pierre, 12, 62, 230–31, 240, 251
Bourne, Randolph, 99, 100–101, 103–7, 120, 238, 246–47
Branch, Taylor, 258–59
Brandes, Stuart, 238
Braverman, Harry, 18, 233
bridge builders, 163–64, 226–28, 261–62

Brook Farm, 96
Brooks, Van Wyck, 34, 99, 100–101, 105–7, 114, 233, 238, 246–47
Brosio, Richard, 231–32, 240
Brown, David, 18–19, 233–34
Brown, Michael, 255
Brown, Michael Jacoby, 254
Brownson, Orestes, 95–96, 246
Burner, Eric, 195, 257–58
Burrow, Rufus, Jr., 245
Bush, Roderick, 174, 206, 255, 259

Camus, Albert, 194–95
Chambers, Edward T. D., 38, 239, 254
citizenship, 3, 10–13, 16, 31, 46–47, 57, 64–66, 72–73, 76–77, 82, 88, 100, 137–39, 205
Civil Rights Act, 212
civil rights movement, 6, 65, 101, 129, 169, 174, 189–90, 203, 208–16, 223, 226, 251, 256–57, 259
  and freedom songs, 198
Cizek, Franz, 116
Cobb, Stanwood, 236, 248, 258
Coleman, Wanda, 18
collaborative progressivism, 1–6, 10, 27–36, 45–47, 51–52, 56–79, 82–84, 88, 91–94, 97, 106, 113–23, 127–29, 132, 144–46, 150–57, 161, 164, 168, 172–73, 177, 181, 185, 190, 192, 198, 216, 221–22, 229
Collins, Patricia Hill, 235–36, 240, 260
Commons, John R., 53, 239
community organizing, 3, 5–6, 10, 28, 33, 37–38, 42, 67, 71, 136, 148, 161–72, 176–86, 192–93, 196–97, 200–201, 213–14, 221–22, 224–26, 238, 253–54, 256–61
  and Association of Community Organizations for Reform Now (ACORN), 172
  and Direct Action Research and Training (DART), 172
  and door-knocking, 172

  and Faith Based Community Organizing (FBCO), 162, 172, 184
  and Gamaliel Foundation, 172
  and one-on-one interview process, 170–71, 174, 223
  and People Improving Communities through Organizing (PICO), 172
  and youth organizing, 224
company unions, 33
Connor, Celeste, 235, 247
consumerism, 34, 98
Counts, George S., 31, 237, 239
Cremin, Lawrence, 107, 124, 248–49
  and *Transformation of the School*, 124, 248–49
critical race theory, 47
Croall, Jonathan, 251
Croteau, David, 239
Curtis, Susan, 233, 236–37, 242, 248

Dahl, Robert, 71, 173, 243–55
Dalton, Thomas, 103, 106, 247–49
Davis, Mike, 233
Day, Dorothy, 93
Deacons for Defense, 191, 213–17, 258–60
democratic education, 4, 11, 41, 46, 52, 54, 92, 113, 115, 124, 150, 155
democratic society, 4, 5, 11, 36, 58, 71–72, 74, 79, 88, 92, 94, 106, 112, 122, 144, 146, 150–51, 182, 221, 227, 242
Dennison, George, 250–51
Dewey, Evelyn, 91, 112, 245
Dewey, John, 1–6, 10–11, 27, 31–36, 39–40, 45–47, 51–88, 91–94, 97, 100–108, 111–46, 149–57, 164, 167, 171–74, 177, 181–82, 185, 190–95, 198–99, 207, 221, 230–31, 237–38, 240–45, 247–52
  on aesthetic symbols, 63–65
  and *Art as Experience*, 117, 122, 242

and *Child and the Curriculum*, 113, 250
and *Common Faith, A*, 86, 244
and *Experience and Education (E & E)*, 92, 117–20, 124, 132, 135, 249, 252
and *Experience and Nature*, 104
and *Freedom and Culture*, 81, 244
and *Human Nature and Conduct*, 132–36, 241, 250, 252
and *Individualism, Old and New*, 106, 117
and *Public and Its Problems, The*, 63, 72–73, 77–83, 85–86, 102, 137, 150, 155, 172, 181, 241, 242, 243
on scientific symbols, 63–64
Dews, C. L. Barney, 260
Dumenil, Lynn, 246–47
Durkheim, Emile, 12, 230, 236
Dyson, Michael, 189, 215, 256, 259–60

Edwards, Anna Camp, 58–60, 69, 82, 237, 241–44
Ellis, Godfrey J., 235
Elshtain, Jean Bethke, 237, 241
Emerson, Ralph Waldo, 95–96, 117, 238, 245, 249
Epstein, Barbara Leslie, 154, 253, 255
Erenreich, Barbara, 236
Estes, Steve, 259

Fairclough, Adam, 258
Fantasia, Rick, 231, 238
Farmer, James, 208–9
Farrell, James, 129, 245, 250
Fass, Paula, 99, 246
Fay, Elizabeth A., 234, 260
Federation Solution, 33
Feinberg, Walter, 237
Fellowship of Reconciliation, 208
feminism, 173
field, 12, 62, 231
Fine, Michelle, 240
Fink, Leon, 9

formal, non-bureaucratic structures, 169, 172, 175, 203
Foucault, Michael, 138, 140, 251
Frank, Waldo, 34, 99, 100–108, 117, 238, 245, 246–47, 256
Freeman, Jo, 255
free schools movement, 10, 35, 91–93, 112, 125, 127–32, 135, 141–42, 146–56, 238, 250
Freud, Sigmund, 98, 100, 108, 118, 120, 127, 130, 133
Friedenberg, Edgar, 140, 153, 251

Gamson, William A., 257
Gandhi, 208–9, 259
Ganz, Marshall, 167
Garrison, James, 241
Gee, James Paul, 230, 233–36
Giddens, Anthony, 230
Glazer, Nathan, 251
Goldman, Emma, 93, 100
Goodman, Paul, 4, 34–35, 127–43, 146–52, 155–56, 238, 250–53
and *Gestalt Therapy*, 135–36, 139, 143–44, 151, 251–53
and *Growing Up Absurd*, 130–31, 140, 238, 251, 253
Gorman, Thomas, 24–27, 235, 236
Gouldner, Alvin Ward, 12, 231
Gramsci, Antonio, 225, 260
Granger, David A., 238, 249
Great Community, The, 74, 77–83, 86–88, 151, 181
Great Depression, the, 16
Great Society, The, 73–74
Greene, Maxine, 87, 190, 244, 256
Gura, Philip F., 96, 246
Gutierrez, Kris, 250

habits, 55–58, 66, 68, 75, 77, 87, 133, 142, 143, 181, 246
habitus, 12, 62
Hart, Betty, 20, 233–34, 238, 250
Hart, Stephen, 239

Harvey, David, 74, 234, 243
Hauser, Caroline, 119, 248–49
Heath, Shirley Brice, 235, 236, 247
Hefferline, Ralph, 135, 250, 252
herd psychology, 101–2, 107, 114–15, 120, 122, 127
higher education, 4, 21, 27, 44–46, 53, 153, 225
Hill, Lance, 210–16, 231–32, 235–38, 246, 254, 258–60
Hillsviewers Against Toxics (HAT), 169
Hoffman, Nicolas von, 165, 174, 255
Holt, John, 127–28
Horton, Myles, 176, 192–93, 256
 and Highlander Center, 176, 192–93, 198, 256
Horwitt, Sanford, 254, 257
Hull, Glynda, 32, 233–35
human nature, 2, 87, 124, 132, 136, 140, 147, 179

impulses, 47, 133–36, 139, 140, 147, 252
industrialization, 13–15, 95

Jailer, Todd, 22
Jung, Carl, 100, 108

Kahn, Si, 254
Kahne, Joseph, 230, 255
Kane, John, 255
Kanigel, Robert, 232, 237
Kanter, Rosabeth Moss, 232–33
Kapur, Sudarshan, 259
Kaufman, Peter, 26, 236
Kennedy, Gail, 86, 212, 244, 247
King, Martin Luther, Jr., 93, 174, 189–93, 197–98, 201–16, 223, 256, 258–59
Kliebard, Herbert M., 45, 240
Kloppenberg, James T., 237
Knight Abowitz, Kathleen, 240
Kohn, Melvin L., 234–35
Kozol, Jonathan, 152–53, 253

Kropotkin, Peter, 93, 151
Ku Klux Klan, 211, 213

Labaree, David F., 11, 46, 230, 240, 253
Laboratory School, 10, 31, 52–71, 82, 110–15, 121–23, 132, 135, 142, 150, 177, 237, 241, 244
 and clubhouse of, 59, 60, 64–65, 84–85
labor unions, 3–5, 10, 13–18, 22–24, 28, 32–37, 40–43, 68, 82, 96, 100, 137, 161–66, 171, 204, 208, 216, 223, 225–26, 233, 236, 253
Laclau, Ernesto, 75, 243
Lagemann, Ellen Condliffe, 240
Lamont, Michele, 21–22, 27, 231, 234, 236
Lareau, Annette, 22–23, 26, 232–36, 261
Law, Eric H. F., 184–85, 230, 239, 256, 260
Lawson, James, 209
leadership, 37–38, 59, 67–68, 94, 152, 162–80, 184, 190–91, 193–208, 212–15, 225
Leidner, Robin, 233, 235
Leondar-Wright, Betsy, 231, 234–35, 239
Leuchtenberg, Walter, 97
Lichterman, Paul, 5, 39, 42–43, 93, 153–54, 169, 173, 199, 227, 230, 239, 245, 253–55, 261
Lippmann, Walter, 30, 39, 71–73, 76, 79, 82–84, 91, 97, 237, 243
 and *Phantom Public*, 72
 and *Public Opinion*, 72, 237, 243
Listen, Inc., 260
Livingstone, David W., 235
Löwy, Michael, 246
Lubrano, Alfred, 23–24, 27, 231, 232, 234–36, 260
Lucey, Helen, 26, 236

Mahoney, Thomas, 232
Manicas, Peter T., 235

Mansbridge, Jane, 71, 81–82, 240, 243–44
manual labor, 13–14, 18, 53, 194
Marable, Manning, 216
March on Washington Movement (MOWM), 208
Martin, Jay, 52, 241
Marx, Karl, 12, 230–31
Mass meetings, 204
Maxcy, Spencer J., 242
Mayhew, Katherine Camp, 58–60, 69, 82, 237, 241–44
McGerr, Michael, 9, 17, 66, 229–30, 233, 236–37, 242
Meiksins, Peter, 233
middle class, 3–6, 9–10, 12–47, 52–53, 56, 62, 66, 69, 79, 92–95, 98–100, 112, 117, 127, 128–29, 139–41, 152–56, 161–66, 169, 173, 178, 183, 184–86, 190–94, 198–206, 209, 214–16, 221–27, 234, 237, 251, 259
Miller, James, 152, 253
Miller, Peggy, 23, 235
Miller, Ron, 151, 235, 238, 250, 253, 255
Mills, C. Wright, 67–68, 128, 242
Milwaukee Inner-city Congregations Allied for Hope (MICAH), 162
Miroff, 173
Montesquieu, 71
Montgomery, AL, 174, 204, 208–10, 213–14
Montgomery, David, 15, 233, 238
Moore, Amzie, 197
Morris, Aldon D., 169, 203, 207, 209, 257, 258, 259
Moses, Robert, 194–201, 207, 212, 257, 258
Mother Jones (Mary Harris Jones), 33, 225, 238
Mumford, Lewis, 100, 103–7, 114, 121, 127, 238, 246–47
  and *Golden Day, The*, 103, 105–7, 247

National Association for the Advancement of Colored People (NAACP), 193, 196, 203
Native Americans, 178
natural consequences, 148
Naumburg, Margaret, 4, 35, 107–24, 128, 177, 221, 238, 248–49
  and *Child and the World, The*, 109, 113, 118, 238, 248
Neill, A. S., 35, 127–33, 136, 141–43, 146–51, 238, 250–52
  and *Free Child, The*, 250, 252
  and *Freedom and Not License*, 128
  and *Summerhill*, 128, 131, 149, 238, 250–52
neuroses, 139–41, 144, 149
Newman, Katherine, 235
Newman, Lance, 246
New Republic, 104–5, 116
1960s, 2, 10–11, 27, 33–35, 91–93, 96–100, 112, 123–30, 137, 140–41, 151–55, 165, 173, 189–92, 201, 212–14, 217, 238, 250–51, 255, 257
nineteenth century, 1, 3, 9–10, 13–16, 19, 23, 28, 44, 93–97
1920s, 2, 10, 27, 33, 34–35, 63, 72, 91–135, 141, 154–55, 229–30, 237–38, 242, 246–47
Noddings, Nel, 4, 11, 108, 156, 229–30
nonviolence, 164, 182, 189, 208–15, 259
Novalis, 95

Ogbar, Jeffrey O. G., 259
Ogren, Christine A., 240
organic intellectuals, 225–27, 261

paradox of size, 72, 80, 88
Pascarella, Ernest T., 234
Payne, Charles, 189, 198, 199, 258
Peggy Miller, 23
performance artist, 142, 144, 154
Perls, Frederick S., 135, 250
Perlstein, Daniel, 189, 256

personalist progressivism, 1–6, 10, 27–29, 33–36, 40, 45–47, 59, 63, 80, 91–96, 99–108, 111–41, 144–45, 151–57, 161, 165, 168, 185, 191–95, 199, 209, 216, 221–22, 238, 245, 248–49
  and nostalgia, 35, 140
Phillips, Anne, 223, 244
Pinkard, Terry P., 246
Piven, Francis Fox, 257
Plato, 71
political theory, 47
Polletta, Francesca, 257–58
popular education, 175–76, 192
poverty, 1, 2, 12, 16, 22, 28, 32, 34–38, 52, 66, 88, 96–99, 140–41, 152–53, 163–65, 166, 170, 178, 182, 183, 185, 197, 200, 205, 209, 222, 223, 226, 250, 253, 254
pragmatism, 57, 100–101, 103–7, 156, 195, 197, 201
Pratt, Caroline, 4, 35, 107–13, 116–21, 124, 128, 131, 238, 248, 249
progressive education, 62, 92, 107, 113, 115, 118, 124, 127, 131, 138, 155

racism, 3, 10, 16, 66–67, 93, 199, 201, 208, 209
Randolph, A. Philip, 99, 208, 225, 238, 246, 247
Ransby, Barbara, 193, 196, 256–58
Rayback, Joseph George, 233, 236
Reitzes, Donald C., 162, 238, 253, 254
Ripley, George, 96
Robertson, Emily, 242
Rockefeller, Stephen C., 86, 242, 244
romanticism, 2, 33, 93, 95–96, 100–101, 117–18, 127, 129
Rose, Fred, 5, 24, 39, 40–43, 163, 185, 226–27, 230, 235–36, 239, 253–54, 261
Ross, Dorothy, 239
Roszak, Theodore, 130, 151, 155, 251, 253

Rugg, Harold, 107, 111, 119, 245–46, 248, 249
  and *Child-Centered School, The*, 111, 248, 249
Ruscio, Kenneth Patrick, 255
Rustin, Bayard, 209

Saito, Naoko, 238, 249
Sales, William W., 174, 206, 259
Sanders, Lynn, 230, 240
Sassoon, Anne Showstack, 260
Schmidt, Bernard, 245
schooling, 2, 4–6, 10–11, 15, 22, 25–26, 35, 44–47, 52–62, 65, 69, 82, 91–92, 96, 107–24, 127–57, 162, 186, 189, 194, 216, 224–25, 232, 250, 251, 260
schools of education, 11, 44–47
Schultz, Brian, 260
Schutz, Aaron, 229–30, 233, 235, 240, 245, 251, 253–54, 256, 260–61
science, 4, 18, 29, 30, 44–45, 53–64, 69, 73–79, 83, 86–87, 92, 97, 103–5, 112, 114, 118, 128, 132, 135, 142, 144, 146, 154, 207
Scott, James C., 237
Sellers, Cleveland, 200
Semel, Susan F., 245
Sennett, Richard, 236
Shirley, Dennis, 235, 239
Shuttlesworth, Fred, 197
Smiley, Glenn, 209–10
social question, 34, 129
social reconstructionists, 31, 237
solidarity, 3, 5, 10, 16, 17, 25–28, 33, 36–39, 41–43, 46, 57, 64, 67, 81, 95, 159, 161, 166, 174, 185–86, 191, 198–200, 205, 229, 231, 238
Southern Christian Leadership Conference (SCLC), 190–94, 198, 201–4, 207, 210–17, 258
Stall, Susan, 229
Staples, Lee, 254
Stembridge, Jane, 194, 196

Stoecker, Randy, 229
Stoehr, Taylor, 136, 250, 252, 253
Storytelling, 235
Stovall, David, 260
Stromquist, Shelton, 65–66, 230, 236–37, 242
Student Nonviolent Coordinating Committee (SNCC), 173, 191–208, 212–17, 257, 259
Students for a Democratic Society (SDS), 128, 152, 154–55, 173, 253
Education Research and Action Project (ERAP), 152, 154, 253
Swarts, Heidi J., 162, 253–54, 257
Swartz, David, 230–31, 240

Taylor, Frederick Winslow, 29, 39, 56, 136, 232, 237, 250, 253, 258
Taylorism, 238
teachers, 11, 26, 44–46, 53–55, 58–61, 68–69, 82–83, 92, 96, 108–16, 121, 124, 131–32, 135, 142, 149, 150, 156, 177, 224, 242
technocracy, 100, 129, 138–41, 150–51, 165
Thoreau, Henry, 33, 95–96, 246
Tokarczyk, Michelle M., 234, 260
Trachman, Matthew L., 255
Trachtenberg, Alan, 231–33, 236–37
transcendentalism, 33, 95–96, 117, 238
Trotha, Trutz von, 23, 234–36
Tudge, Jonathan, 233–34

Tyack, David B., 45, 240

upper class, 13, 33, 44, 52, 66, 141, 232

Veysey, Lawrence R., 239

Wade, Rahima, 260
Walkerdine, Valerie, 26, 236
Warren, Mark R., 239, 254
Weber, Max, 12, 230–31
Weltman, Burton, 250, 252–53
West, Cornel, 237, 242, 245, 260
Westbrook, Robert, 52, 82, 85, 87, 92, 123, 237, 241, 243–45, 247–48
Westheimer, Joel, 230, 255
Whitman, Walt, 33, 93, 95–96, 242
Wiebe, Robert H., 29, 237
Willis, Paul, 235
Wilmore, Gayraud, 202, 258
working class, 3–6, 9–47, 52–55, 62, 66–67, 73, 88, 95–97, 100, 129, 137, 140–41, 161–71, 178, 182–86, 191–93, 196–97, 200–207, 211, 214–17, 222–27, 230, 234–35, 237, 240, 246, 256
working-class scholars, 225
Wright, Eric Olin, 67, 128, 230, 242

X, Malcolm, 174, 206, 259

Zolo, Danilo, 243

GPSR Compliance

The European Union's (EU) General Product Safety Regulation (GPSR) is a set of rules that requires consumer products to be safe and our obligations to ensure this.

If you have any concerns about our products, you can contact us on

ProductSafety@springernature.com

In case Publisher is established outside the EU, the EU authorized representative is:

Springer Nature Customer Service Center GmbH
Europaplatz 3
69115 Heidelberg, Germany

www.ingramcontent.com/pod-product-compliance
Lightning Source LLC
LaVergne TN
LVHW011808060526
838200LV00053B/3707